Optical Engineering Fundamentals

Second Edition

Tutorial Texts Series

Optical Engineering Fundamentals

Second Edition

Bruce H. Walker

Tutorial Texts in Optical Engineering
Volume TT82

SPIE PRESS

Bellingham, Washington USA

Library of Congress Cataloging-in-Publication Data

Walker, Bruce H.
  Optical engineering fundamentals / Bruce Walker. -- 2nd ed.
      p. cm. -- (SPIE tutorial texts)
  Includes bibliographical references and index.
  ISBN 978-0-8194-7540-4
  1. Optics. 2. Lenses. 3. Optical instruments. I. Title.
  TA1520.W35 2008
  681'.4--dc22

                                                                2008047526

Published by

SPIE
P.O. Box 10
Bellingham, Washington 98227-0010 USA
Phone: +1 360 676 3290
Fax: +1 360 647 1445
Email: books@spie.org
Web: http://spie.org

Introduction to the Series

Since its inception in 1989, the Tutorial Texts (TT) series has grown to more than 80 titles covering many diverse fields of science and engineering. The initial idea for the series was to make material presented in SPIE short courses available to those who could not attend and to provide a reference text for those who could. Thus, many of the texts in this series are generated by augmenting course notes with descriptive text that further illuminates the subject. In this way, the TT becomes an excellent stand-alone reference that finds a much wider audience than only short course attendees.

Tutorial Texts have grown in popularity and in the scope of material covered since 1989. They no longer necessarily stem from short courses; rather, they are often generated by experts in the field. They are popular because they provide a ready reference to those wishing to learn about emerging technologies or the latest information within their field. The topics within the series have grown from the initial areas of geometrical optics, optical detectors, and image processing to include the emerging fields of nanotechnology, biomedical optics, fiber optics, and laser technologies. Authors contributing to the TT series are instructed to provide introductory material so that those new to the field may use the book as a starting point to get a basic grasp of the material. It is hoped that some readers may develop sufficient interest to take a short course by the author or pursue further research in more advanced books to delve deeper into the subject.

The books in this series are distinguished from other technical monographs and textbooks in the way in which the material is presented. In keeping with the tutorial nature of the series, there is an emphasis on the use of graphical and illustrative material to better elucidate basic and advanced concepts. There is also heavy use of tabular reference data and numerous examples to further explain the concepts presented. The publishing time for the books is kept to a minimum so that the books will be as timely and up-to-date as possible. Furthermore, these introductory books are competitively priced compared to more traditional books on the same subject.

When a proposal for a text is received, each proposal is evaluated to determine the relevance of the proposed topic. This initial reviewing process has been very helpful to authors in identifying, early in the writing process, the need for additional material or other changes in approach that would serve to strengthen the text. Once a manuscript is completed, it is peer reviewed to ensure that chapters communicate accurately the essential ingredients of the science and technologies under discussion.

It is my goal to maintain the style and quality of books in the series and to further expand the topic areas to include new emerging fields as they become of interest to our reading audience.

James A. Harrington
Rutgers University

Contents

Preface to the First Edition

The concept of this book, along with many of the ideas and much of the material contained here, has evolved over the last 25 years. During that time, while working in the field of optical engineering and lens design, I have frequently been called on to describe and explain certain optical theories or phenomena to coworkers or readers of trade publications. In responding, my goal has always been to reduce these explanations to the most basic technical level, one that is easily comprehensible. The value of this book, then, is not in bringing forth new information never previously available to the reader. Rather, it is a carefully thought-out selection of material that I feel will be of maximum interest and value to the reader. The material will be presented in a form that can be easily understood in the absence of complex theories of mathematics and physics.

The field of optical engineering and the subject of optics in general are not merely interesting, but often quite fascinating. In order to be a good optical engineer, one must have a sincere interest and curiosity about the subject. This must then be supplemented by a fundamental knowledge of just a few very basic principles that will allow that curiosity to be satisfied. This book is designed to assist the student or worker who is interested and involved in the field of optics to obtain a better understanding of those basic principles and to prepare the reader for the more complex topics that will be encountered using more advanced, specialized textbooks and reference material.

Thanks to the extraordinary nature of the human visual system and the many wonders of the world in which we live, hardly a day passes that we are not exposed to the science of optics, albeit often without a full understanding of what we are experiencing. Consider modern technology like home television systems that include projection TVs, laser disks, and compact camcorders. In the case of home audio systems, consider the compact laser disk revolution that, in just a few short years, has made the traditional phonograph record essentially obsolete. Consider also the optical technology that has been demonstrated by the space and defense industries in recent years. We have been privileged to view pictures from space and from distant planets. Military conflicts have been decided by the use of "smart" bombs, laser sights, and head-up displays . . . all sophisticated products of today's optical engineer.

These are wonders of our own making. We need only observe a colorful sunset, or view a rainbow, to witness some of the many optical wonders of nature. Of course we would not observe this or anything else were it not for the most amazing of nature's optical systems: the human eye and its associated physiological components. It is my sincere hope that this book will, in some small way, make it possible for readers to understand and appreciate many of these things, so that they might feel more a part of all that goes on around them, especially as it relates to the science of optics and the field of optical engineering.

If you work in the field of optics, I'm certain you will find that a better understanding of the subject will serve to improve your basic skills. In addition, I believe it will also enhance your enjoyment of not just your work, but hopefully of life in general. Equally important, it is this understanding that makes it possible for us to share these experiences with others.

One goal of this book is to demonstrate that there are many aspects of the science of optics and the field of optical engineering that are not that difficult to understand. Until now, the majority of texts and other publications in this field have assumed a certain level of understanding and proceeded from that point. In some 25 years of work as an optical engineer, which has included the preparation and presentation of numerous technical papers and articles for a variety of publications, I have encountered a very real need, and a sincere desire on the part of many, to obtain that information required in order to reach that assumed level of understanding. This book is presented in the belief that it will serve to meet that need and to quench that desire.

Bruce H. Walker
November 1994

Preface to the Second Edition

The first edition of *Optical Engineering Fundamentals* (*OEF*) reflected my experience and exposure to the history of optical engineering as I witnessed it during 25 years of work within the optics industry. All of these work assignments involved the usual team structures, where the range of duties and responsibilities of the optical engineer were quite limited.

In the years 1992 through 2008, I continued to work actively as an independent consultant in the fields of optical engineering, lens design, and optomechanical systems design. This second edition of *OEF* reflects much of what I have learned during this more recent personal work experience, with emphasis on the application of state-of-the-art computer software to generate solutions to related lens design and optimization problems. While all lens design examples presented involve the use of the OSLO software package (Lambda Research, Littleton, MA), it must be made clear that competitive software packages, such as ZEMAX, Code V, and others can be applied to generate similar results.

All of this consulting work has required a more complete and thorough understanding of the optical engineering, design, and manufacturing process. Many of the changes and additions contained in this second edition reflect that increased scope of work.

Another area that has witnessed significant changes in recent years includes computer systems and software for optical design and analysis. This second edition of *OEF* contains a chapter dealing with the OSLO-EDU software package. It is hoped that all interested readers, regardless of their professional responsibilities, will find that being familiar with tools like OSLO-EDU will permit them to do a better job.

The early chapters of *OEF* deal with basic concepts that remain unchanged. This edition will expand on those concepts and improve on the effectiveness of the presentations.

Subsequent portions of this edition deal with specific optical components, instruments, and systems. These topics are updated to reflect recent developments in those areas, in particular, the development of electronic sensors and how they impact the work of the optical engineer.

Acknowledgments

One of the many advantages of a career in the field of optical engineering is the assurance that one will encounter a wide variety of very talented and interesting people in the course of one's work. As a young fledgling engineer at General Electric in the early 1960s, my interest in optics and lens design was sparked and nurtured by three such individuals. For their early and sustained inspiration and support, I would like to thank Dr. Jack Mauro, Mr. Bob Sparling, and Mr. Don Kienholz.

I would like to dedicate this second edition of *Optical Engineering Fundamentals* to the memory of Warren J. Smith, who passed away earlier this year. In 1968, while at General Electric Co., I completed an in-house course on basic optics. The text used was *Modern Optical Engineering* by Warren J. Smith. This book was my introduction to optical engineering, and it would have a profound impact on the remainder of my professional life. Later that year I submitted my first article for publication in *Optical Spectra*. While I was already quite proud of being published, that pride increased exponentially when I received a personal note from Warren complimenting my work. Twenty-five years later, while searching (without much luck) for a publisher of *Optical Engineering Fundamentals*, I contacted Warren for advice. He quickly reviewed my proposal and put me on the fast track with the people at McGraw-Hill, where my book was published in 1994. Warren was a great engineer, a good friend, and a wonderful person. We will all miss him.

In addition, I would like to thank the staff at SPIE, especially those in the publications department, for their enthusiastic and effective support over the years.

Bruce H. Walker
Walker Associates
December 2008

Optical Engineering Fundamentals

Second Edition

Chapter 1
Introduction

Optical Engineering Fundamentals is intended to be a "bridge" book, i.e., a book that will carry the reader who is interested in the field of optical engineering from a position of curiosity, confusion, and apprehension, to one of understanding, comfort, and appreciation. While this book alone will not make you an optical engineer, it is hoped that it will enable you to communicate and work closely and effectively with the optical engineer. Should it be your goal to eventually work in the field of optical engineering, this book will provide major assistance in two ways. First, it will provide a basic understanding and appreciation of many fundamental optical principles, thus preparing you to get the most out of the many fine advanced optics texts and courses of study that are available. Second, as you work into the field, this book will serve as a convenient source of valuable reference material.

How many of us have said to ourselves (or heard it said), "I don't know anything about optics, but…" and then the speaker has proceeded to conclusively prove that point? It is this author's fondest hope that this book will render that phrase obsolete for its readers. Frequently, an optics subject is brought up in a discussion, couched in terms that are esoteric beyond the understanding of the layperson or novice. This book will provide the reader with a level of knowledge that will assure a degree of comfort in discussing the basics of such topics and exploring their more advanced aspects, as required.

For many years the job descriptions for the optical engineer and the lens designer were considered to be quite separate and distinct. Because the science of lens design was so unique, and because the calculations involved were so complex and time consuming, the lens designer rarely had the time or inclination to venture into the more general fields that were the responsibility of the optical engineer. Likewise, the successful optical engineer has traditionally been more than willing to leave the tedious task of lens design to the specialist. However, recent developments in modern technology have led to a significant realignment in these areas.

With today's modern personal computer systems and easy-to-use (read "fun") software packages dedicated to lens design and optical analysis, it behooves today's optical engineer to develop a certain level of competence in this area. While the intricacies of starting lens-type selection and detailed lens optimization are still best left to the lens designer, the ability to utilize a modern optics program for performing basic optical system analysis will make one a much more effective optical engineer. Likewise, the speed with which lens design tasks can now be accomplished allows the lens designer to diversify his or her activities into some of the more general areas of optical engineering. The two classifications that once were quite distinct and clearly defined are gradually blending together into a single position, often labeled *optical designer.* This is good for the optics industry, in that there has always been a shortage of qualified personnel in both areas. More important, it is very good for individuals who select optical design as a career path in that it broadens the scope of their work, making it more interesting, challenging, and rewarding.

It is believed that anyone working in or dealing closely with the optics industry will find this book both interesting and useful. A first reading will create a sense of comfort with the subject matter not obtainable from previously available material. As a teaching tool, this book will prove most valuable in exposing the student to the intriguing science of optics and optical engineering without the complications of advanced physics and mathematical theory. Subsequently, the formulas, illustrations, and tables of information contained within the book will serve as valuable reference material for all working within or in close proximity to the field of optical engineering.

The book begins with a historical review, covering some of the more interesting people and events that have been involved in the study and application of the science of optics over the years. While not intended to be comprehensive, this historical information will be of interest in terms of the relationship between modern optical engineering and the events and people of the past.

Similarly, an early chapter in the book deals with the age-old question, "What is light?" While a definitive answer is not forthcoming, the presentation should result in a certain level of comfort with the topic that will allow the reader more confidently and intelligently to join with the likes of Aristotle, Ptolemy, and Galileo in pondering the ultimate answer to this most basic question. The relationship between light waves and light rays is then presented. The laws of reflection and refraction and their relationship to basic optical components are introduced, as is the subject of dispersion. The topic of diffraction is also touched on, to the extent that it affects the performance and image quality of many typical optical systems.

The topic of thin-lens theory is developed and discussed, along with several examples of how it can be applied to real-world problem solving. This topic is followed by an introduction to a basic computer software package, designed to aid the optical engineer in understanding and evaluating the optical system. Several examples of thin-lens theory application are included. For the transition from a preliminary design to a final configuration, a basic understanding of the primary aberrations of a lens system is most helpful. Coverage of this topic includes a basic description of these aberrations with a discussion of their origins and impact on performance and image quality of the lens system.

Having established this foundation, the book then devotes several chapters to a discussion of the design, function, and application of basic optical devices and instruments. Details of a variety of typical optical components are discussed, including the manufacturing processes employed and the determination of specifications and tolerances. The characteristics, behavior, and relative strengths and weaknesses of optical materials are covered. Included among these are optical glass, metal optics, coatings, and special materials for the ultraviolet and infrared portions of the spectrum. The special considerations of optical engineering as it relates to visual systems are covered, including a comprehensive review of the general characteristics, configurations, and functions of the human eye. A separate chapter is dedicated to modern lens design methods, including a brief historical review of computing techniques and the modern computer as it relates to the history of lens design. Finally, several of the more interesting optical devices and phenomena encountered in day-to-day life are covered.

If the reader comes away with one predominant sense after reading this book, I hope it will be an awareness of the many ways in which optics and optical engineering are interrelated with our everyday lives. Every book has its purpose; in this case that purpose is to present some of the more unique, interesting, and useful aspects of optics and optical engineering in a format that will be enjoyable, useful, effective, and entertaining . . . rather than intimidating.

Chapter 2
Historical Review

2.1 Definition of Optical Engineering

By definition, *optics* is the scientific study of light and vision, chiefly of the generation, propagation, manipulation, and detection of electromagnetic radiation having wavelengths greater than x rays and shorter than microwaves. The term *optical* applies to anything of or pertaining to optics. By further definition then, *optical engineering* would involve the application of scientific optical principles to practical ends, such as the design, construction, and operation of useful optical instruments, equipment, and systems. This chapter will deal with some of the people, places, and events that, over the years, have contributed significantly to the history of optical engineering.

2.2 Ancient History

The earliest indications of some knowledge and application of optical principles appeared nearly 4000 years ago when two unrelated massive stone structures, Stonehenge and the Pyramid of Cheops, were constructed. In both cases the orientation of these structures is found to be tied closely to the relationship between the earth and the sun. Familiarity with the cyclical relationship of the seasons, along with the knowledge that light travels in straight lines, would have been required for the resulting arrangements to occur. The precise orientation of these structures relative to the compass and the calendar may be interpreted as an early demonstration of basic optical engineering principles.

Early historical records of Plato and Aristotle (ca. 350 BC) reveal one of the first debates over the exact nature of light. While Plato taught that vision was accomplished by the expulsion of ocular beams from the eyes, his student Aristotle rejected that theory, arguing (more correctly) that vision arises when particles emitted from the object enter the pupil of the eye.

While Euclid (ca. 300 BC) is best known for his study and writings dealing with geometry, he contributed a great deal to the science of optics through his book titled *Optics*. While erroneously accepting Plato's theory of ocular beams,

Euclid did, among other things, correctly describe the formation of images by spherical and parabolic mirrors.

Knowledge of geometry and knowledge of optics were elegantly combined in an experiment to determine the circumference of the earth by a Greek scholar named Eratosthenes. Around the year 230 BC, Eratosthenes was serving as the head of the library in Alexandria, Egypt. There, he read that at noon on the day of the summer solstice each year, in the city of Syene (now Aswan), 800 km to the south, light from the sun was seen to pass directly down into a deep well and then to reflect directly back to the eye from the horizontal surface of the water deep within the well. Eratosthenes noted that at the same time, on the same day, a local vertical stone tower in Alexandria cast a shadow of easily measured length. Using this information he was able to conclude that the angle between the tower and the incident sunlight was 7.2 degrees. He assumed that extensions of the well in Syene and the vertical tower in Alexandria would intersect at the earth's center (see Fig. 2.1). Using this information he was able to calculate the earth's circumference to be 50 times the distance from Syene to Alexandria (50 × 800 = 40,000 km). Quite remarkably, this agrees with today's accepted value within 0.1%!

Ptolemy (ca. AD 150) was yet another great scientist from Alexandria. In his book *Optics*, Ptolemy carries on the ocular beam theory, while also describing correctly the law of reflection (the incident light ray, the reflected ray, and the normal to the reflecting surface all lie in a common plane, with the angle of incidence being equal to the angle of reflection). Ptolemy also studied refraction and published a similar theory, but he failed to establish the exact relationship between the angle of incidence and the angle of refraction.

2.3 Medieval Optics

Alhazen (ca. 1000) was a preeminent Arabian scholar who examined and explained the anatomy and function of the eye. It is thought that his work may have inspired the invention of spectacles. Alhazen is also credited by some with the invention of the camera obscura. There is a distinct functional similarity between the human eye and the camera obscura (Fig. 2.2); this fact may have been instrumental in the work done by Alhazen.

During this time period there was a revival in interest in optics in the west. Vitello of Poland wrote a treatise on optics around 1270. While based in large part on the works of Ptolemy and Alhazen, Vitello's work did much to stir the interest of other medieval scientists. Around 1260, Roger Bacon (1214–1292) was studying and experimenting with lenses to improve vision. Whether Bacon deserves credit for the invention of spectacles remains a subject of some debate. It is certain that spectacles were in use by the time of his death.

Led by the likes of Chaucer, Gutenberg, and Columbus, these years were witness to significant achievements in many fields other than optics. In 1550 Francesco Maurolico of Naples began a systematic study of prisms, spherical mirrors, and the human eye. He described the correction of nearsightedness with

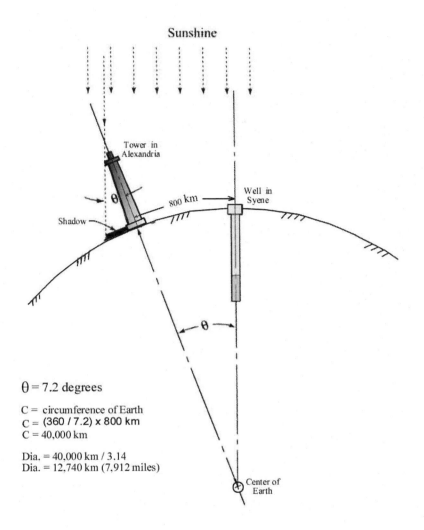

Figure 2.1 Around 230 BC, knowledge of geometry and optics were cleverly combined by the Greek scholar Eratosthenes to estimate the diameter of the earth. Knowing the distance from Syene to Alexandria, and the relationship of the tower to its shadow when the sun was directly overhead at the well, he concluded that the earth's diameter was approximately 12,740 km.

a concave (negative) lens and farsightedness with a convex (positive) lens. As had others, Maurolico attempted to derive the law of refraction, but he was not successful. In 1589 Giambattista della Porto, also of Naples, published a treatise on lenses that contained construction information for a telescope with a positive converging objective lens and a negative diverging eye lens. It has been written that the first telescopes were the result of fortuitous observations by a Dutch spectacle maker. Johannes Janssen of Holland declared in 1634 that his father had "made the first telescope amongst us in 1604, after a model of an Italian one,

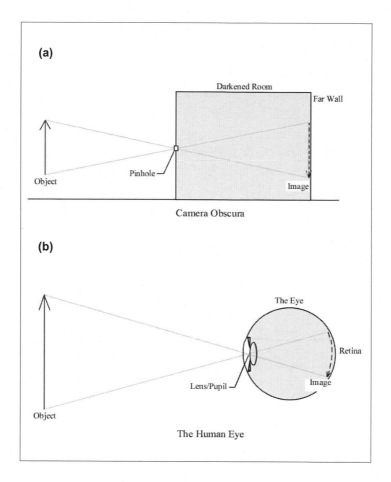

Figure 2.2 The camera obscura (a) and the human eye (b) were both studied by the Arabian scholar Alhazen, ca. 1000. The optical similarity of the two is clearly shown in this illustration.

on which was written anno 1590." Whether the telescope was a Dutch or Italian invention, the Italian physicist Galileo Galilei (1564–1642) learned of the invention from the Dutch sources and was quickly able to duplicate it. In his hands the telescope did much to transform medieval natural philosophy into modern science.

Also instrumental in the transformation of optics into modern science was Johannes Kepler (1571–1630). In 1604 Kepler published a work containing a good approximation of the law of refraction, a section on vision, and a mathematical treatment on those optical devices and systems then known.

2.4 From 1600 to the 1900s

Although Willebrord Snell (1580–1626) is said to have formulated the law of refraction in 1621, he did not publish it, and it did not become known until 1678 when his results were published by Christiaan Huygens (1629–1695). During this time the same law was formulated by Rene Descartes and James Gregory. The honor of precedence was given to Snell and, as a result, the law of refraction carries his name today.

Later in the seventeenth century, a number of experiments were carried out which added greatly to the understanding of light and its behavior. Several experiments by Francesco Grimaldi (1618–1663) demonstrated the phenomenon of diffraction, although many years would pass before the significance of his work was completely understood and appreciated. Sir Isaac Newton (1642–1727) conducted experiments that demonstrated the dispersion of white light by a glass prism. Newton's book on optics, *Opticks*, was first published in 1704 and would remain an influential reference for more than a century.

A large part of the history of optics from 1650 to today is represented by the independent development of two basic optical instruments and one related process: the telescope, the microscope, and photography. It was in 1668 that, while a 26-year-old graduate student at Cambridge, Isaac Newton wished to build a telescope for his own use. After carefully considering the potential problems attached to the use of lenses to build a refracting telescope, Newton opted to build a reflecting telescope with a metal mirror for an objective. While only 33 mm in diameter, with a focal length of just 152 mm, this successful telescope represented a significant accomplishment for the time.

During these years the use of large lenses to produce a refracting telescope objective was frustrated by the lack of understanding of the dispersive properties of optical glass and also by the lack of adequate quality in available glass blanks. It would be some 60 years later, around the year 1730, that an obscure 25-year-old lawyer named Chester Moor Hall (1704–1717) solved the problem of dispersion when he designed and built the first achromatic doublet. This was a two-element lens that combined lead oxide glasses (flints) with the commonly available crowns. Sensing the value of his invention, Hall attempted to keep its form a secret by selecting two different optical shops to work on different halves of his lenses. As it developed, each shop turned out to be busy and passed the job on to someone else. Unfortunately for Hall, each shop chose the same subcontractor and his secret was out.

Still, the glass quality problem limited the diameter of achromatic lenses to just a few inches. In the early 1800s, a number of people were involved in developing new and better methods of glass manufacture to eliminate these quality problems. As a result, around that time (1800) Joseph Fraunhofer (1787–1826) was able to produce achromatic doublets as large as 8.5 in. in diameter.

The mirrors used by Newton and his contemporaries to build telescopes were polished into base-metal substrates that needed to be frequently repolished in order to eliminate tarnishing. In 1850, Justus von Liebig (1803–1873) invented a process for the chemical silvering of glass substrates. After this, nearly all mirrors were made of glass and simply stripped of their silver coating and resilvered when they became tarnished. It wasn't until the 1930s that this process was replaced by the use of deposited aluminum, which essentially eliminated the tarnishing problem. Also in the 1930s, several revolutionary reflecting objective lens designs appeared. The Ritchey-Chretien and Schmidt configurations would greatly increase the limits of image quality and field of view for the reflecting telescope objective.

Paralleling the development of the telescope during these years was the development of the microscope. Devised originally by the Dutch father-son team of Hans and Zacharias Janssen in the late 1500s, the microscope would later be improved upon by many others. The simplest version, containing just a single lens of very short focal length, was described and used extensively by the Dutch biologist van Leeuwenhoek (1632–1703). An early form of the improved compound microscope appears in the publication *Micrographica*, by Robert Hooke (1635–1703) in 1665. It was in the late 1700s that achromatic microscopes were first produced by John Dolland (1706–1761) and others. The problem of combining several corrected achromatic doublets to produce increased magnification was solved in 1830 by an amateur microscopist named Joseph Jackson Lister (1786–1869), father of the famous surgeon Joseph Lord Lister. Around 1850, the Italian optician Giovanni Amici (1786–1868) added a plano-convex lens to the front of the microscope objective, resulting in even greater magnification. In 1879, Ernst Abbe (1840–1905) established the relationship between numerical aperture and resolving power in a microscope objective. Abbe also introduced the concept of homogeneous immersion and later the use of crystalline fluorite for the elimination of secondary color in microscope objectives. In the early 1900s there was a flurry of activity involving the use of ultraviolet illumination in microscopy to achieve the improved resolution that accompanied the reduced wavelength. In 1928 this all became academic with the introduction of the electron microscope, which brought with it a 100× increase in resolving power.

The process of photography was a dream of many during these years, as the images of the camera obscura were observed but could not be recorded. Finally, in 1839, the French artist Louis Daguerre (1787–1851) announced his system of photography, using an iodized polished silver plate, which was later developed with mercury vapor. For a lens with a large flat field, Daguerre turned to his friend Charles Chevalier (1804–1859) who suggested the use of a reversed telescope objective with an aperture stop in front that would limit the lens speed to $f/15$. The need for increased lens speed soon became apparent and in less than a year the 32-year-old mathematics professor at the University of Vienna, Joseph Petzval (1807–1891), designed his famous portrait lens with a speed of $f/3.5$. This new lens design resulted in an increase in image brightness of nearly 20× over that of the lens by Chevalier. From that time, and continuing to this day, the design of lenses has been greatly influenced by the needs and desires of the photographer. The wide angle concentric lens design by Schroeder at Ross Optical in 1887 and the anastigmat by Paul Rudolph (1858–1935) at Ziess in 1900 were monumental lens design achievements. Since that time, while the achievements in photographic lens design have been great in number, they have been more incremental in nature.

In 1781 Sir William Herschel (1738–1822) utilized modern optical technology to discover the planet Uranus, nearly doubling the extent of the known solar system. While conducting further studies involving the sun, Herschel discovered thermal activity in dispersed sunlight beyond the red end of the visible spectrum. Following much debate, Thomas Young (1773–1829) was responsible for experiments that confirmed the existence of infrared (beyond red) energy. Young's work went a long way toward explaining the wave nature of light, including the diffraction effects that had been observed by Grimaldi some 150 years earlier. In addition, as a physician of some note, Young did considerable research and reporting on the mechanism of the human eye.

2.5 Speed of Light

Over the years the history of optics has been tied inexorably to the quest to determine the velocity of the propagation of light within various media. Initially, it was thought that light traveled with infinite speed. As early as the eleventh century it was thought that light did travel at a finite speed, but much too fast to be measured using normal methods. In 1675 the Danish astronomer Olaf Roemer (1644–1710) made the first scientific determination of the speed of light based on observations of the eclipses of the innermost moon of Jupiter. Roemer noted a significant difference in the timing of these eclipses, depending on the relative positions of the Sun, Earth, and Jupiter when the observations were made. Essentially, when the earth was nearest to Jupiter the eclipses would occur several minutes ahead of the predicted time and when the earth was farthest from Jupiter, that eclipse would occur several minutes later than was predicted. While there is no record that Roemer actually did the final calculation, his data would lead to the conclusion that light travels at a speed of 200,000 km/s. Contemporaries of Roemer would modify his findings, including more accurate data on the earth's orbital radius, and arrive at a value close to 300,000 km/s.

Approximately 50 years later, in 1728, the noted British astronomer James Bradley (1693–1762) made an entirely different type of astronomical observation from which he was able to calculate the speed of light. This experiment involved the observation of a star using a telescope with its axis set perpendicular to the plane of the earth's rotation. It was found that in order to compensate for the speed of the incident light, the telescope's axis would have to be tilted through a small angle in the direction in which the earth is traveling. The amount of telescope tilt required allowed Bradley to calculate the speed of light, which he found to be 301,000 km/s.

The first terrestrial measurement of the speed of light was recorded by the French scientist Armand Fizeau (1819–1896) in 1849. Fizeau's experiment is illustrated in Fig. 2.3. A light source was focused through a beamsplitter onto an image plane where a spinning toothed wheel was located. The light passing between teeth of that wheel was then projected to a mirror at a distance of about 8 km, where it was collected and then reflected back to the point of origin. The rotational speed of the wheel was then increased until the returning light was

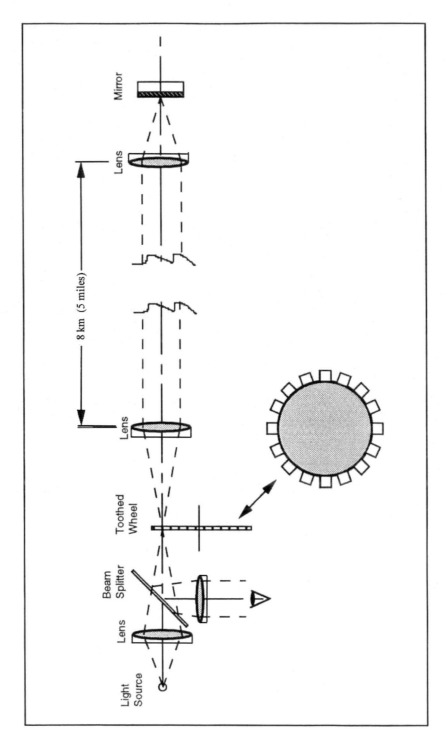

Figure 2.3 The first terrestrial measurement of the speed of light was done by Fizeau in 1849 when he projected a pulsed beam of light onto a distant mirror. Based on the number of teeth and speed of rotation of the toothed wheel, and knowing the distance to the mirror, he was able to calculate a speed of 315,000 km/s.

blocked by the tooth on the wheel just adjacent to the space through which it had passed. Using this data, Fizeau was then able to calculate the speed of light. Limited by the precision of his measurements, Fizeau calculated the speed of light to be 315,000 km/s. Fizeau's experiment was later modified by French physicist Jean Léon Foucault (1819–1868), who replaced the toothed wheel with a rotating mirror. With this new arrangement Foucault determined the speed of light to be 298,000 km/s, much closer to today's accepted value. Foucault was also able to insert a tube filled with water between the rotating mirror and the distant mirror and determine conclusively that the speed of light was reduced when it was traveling through water rather than air. This conclusion went a long way toward disproving the corpuscular theory, which stated that the speed of light in water would be greater than it is in air. The Foucault method was further improved by many, with the most precise measurement being made by Albert A. Michelson (1852–1931). The average result of a large number of measurements done by Michelson was 299,774 km/s. Many aspects of modern technology have been applied to the determination of the speed of light in recent years, yielding a currently accepted value of 299,793 km/s. Finally, it is interesting to note that electromagnetic theory allows the velocity of electromagnetic waves in free space to be predicted, with a resulting value of 299,979 km/s, which is within 0.1% of the most precise measured values.

2.6 Modern Optical Engineering

Optical engineering in the twentieth century has witnessed many sweeping technological advances. A review of developments relating to the photographic lens system will serve well to demonstrate one such area. At the turn of the twentieth century, photography had been in existence for some 50 years, and only a few basic lens types were available to the photographer. A more thorough understanding of lens aberration theory and the mathematical methods of ray tracing would now permit the design of more sophisticated lens forms, such as the Cooke triplet by H. Dennis Taylor (1862–1943) and the Zeiss Tessar lens by Paul Rudolph (1858–1935). Testimony to the quality of these designs is the fact that both remain in common use by photographers today, nearly 100 years later. Another monumental design was the Lee Opic lens, designed by H. W. Lee (1879–1976) in 1920. The Lee Opic is an unsymmetrical double-Gauss lens, derivatives of which have been supplied as the standard lens on many of the 35-mm cameras produced in the latter half of the century. Figure 2.4 illustrates these three classic lens designs.

Starting around 1950, up to and including today, the variable focal length, or zoom lens, has been through a most fruitful period of development. Many of the most significant technological breakthroughs in a number of fields have been involved in the design and production of today's high-performance, compact-zoom lenses for television, motion picture, and still photography.

The development of these high-performance lenses would not have been possible without simultaneous advances in the field of optical glass manufacture.

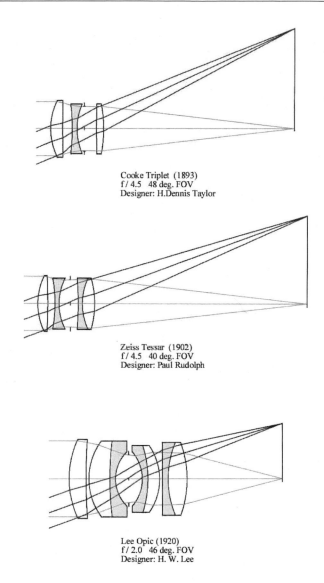

Cooke Triplet (1893)
f / 4.5 48 deg. FOV
Designer: H.Dennis Taylor

Zeiss Tessar (1902)
f / 4.5 40 deg. FOV
Designer: Paul Rudolph

Lee Opic (1920)
f / 2.0 46 deg. FOV
Designer: H. W. Lee

Figure 2.4 Three milestones in the recent history of photographic lens design. Each of these basic lens design forms remains in common use today.

In Europe and Great Britain several precision glass manufacturers were in place at the turn of the century. Soveril, in France, had a history dating back to 1832, while the roots of the Chance-Pilkington glass company in England could be traced back to 1824. In Germany, the Schott Glass Works had begun its work in 1880. In the United States, Bausch and Lomb established a glass manufacturing capability in 1912. This proved to be most fortunate when in 1914 the start of World War I cut off all supplies of optical glass from Europe. At Eastman Kodak a program was begun in 1937 that involved the manufacture of special optical glasses formulated specifically for the purpose of developing improved lens designs. Since the end of World War II in 1945, the Japanese have become major producers of high-quality optical glass. This is just one of the factors that made possible the virtual domination of the 35-mm camera market by Japan in the second half of the twentieth century. The two principal Japanese glass manufacturers today are Ohara and HOYA.

Another area of optical technology that has evolved dramatically during the twentieth century is that of optical coatings to reduce losses due to reflection at air-to-glass surfaces. Early lens designs had just a few air-to-glass interfaces. As a result, the amount of light lost due to surface reflections did not represent a significant problem. As photographic lens designs became more complex, lenses such as the Cooke triplet, with six air-glass surfaces, became common. In this type of lens, surface reflections would reduce overall light transmission to about 70%. While the 30% light loss was important in its own right, a second problem existed in that much of the reflected light would eventually find its way to the image plane in the form of stray light and ghost images, thus greatly reducing final image contrast and clarity.

Perhaps influenced by the fact that these reflections detracted from his otherwise outstanding lens design, H. Dennis Taylor (designer of the Cooke triplet) was instrumental in early studies of antirefection coatings. Taylor observed in 1896 that certain lenses that had been tarnished by environmental exposure exhibited improved light transmission characteristics. Around that same time, while working at Ross Ltd. in England, Frederick Kollmorgen (1871–1950) is said to have observed similar improved light transmission through portions of a magnifying lens that had been accidentally stained with spilled ink. Various methods of chemically coating optics were attempted by Taylor, Kollmorgen, and others with varying degrees of success due to the unpredictable nature of the process. It was in 1935 that Alexander Smakula (1900–1983) of Zeiss invented a process of coating lens surfaces in a vacuum with a thin evaporated layer of a low-index material such as calcium fluoride or magnesium fluoride. The same process was described by John Strong (1905–1992), a researcher at California Institute of Technology, in 1936.

Subsequently, as the evaporated coating process was perfected, it was found that multiple layers of certain materials would further reduce the amount of reflected light. With these multilayer antireflection coatings it is possible today to produce a complex lens assembly with 20 air-to-glass surfaces that will demonstrate essentially the same light transmission characteristics as did an air-spaced doublet with no coatings.

The process of lens design will be touched on here from a historic perspective and discussed in greater detail in a later chapter. The status of lens design in 1840 can be illustrated by the story relating how Joseph Petzval was assigned the help of two artillery corporals and eight gunners, all of whom were "skilled in computing," to aid him in the calculations required to complete a lens design. At the end of six months, this team had completed just two basic lens designs, one of which was the famous Petzval portrait lens.

Lens design procedures remained unchanged into the early part of the twentieth century. Those involved in lens design work employed mathematical formulae for the computation of aberrations and for the exact tracing of rays through lens surfaces. The accuracy required called for the use of seven-place logarithms and for all calculations to be executed with extreme caution. Frequently, as was the case with Petzval, teams of several individuals were assigned to the lens designer to execute these tedious calculations. It was in the 1930s that the early mechanical calculator was developed and this burden was just slightly reduced. Still, the design of a complex lens would involve the work of several people, working over periods of months and often years.

Finally, in the 1950s, the electronic computer appeared on the scene and became available for use in the solution of lens design problems. In the period from 1960 to 2008, progress in the area of computer hardware and software has been nothing short of remarkable. A skilled lens designer, equipped with a modest personal computer and appropriate software, can duplicate the work done in six months by Petzval and his team of corporals and gunners, in a matter of minutes.

Any discussion of optical engineering in the twentieth century would not be complete without mention of the laser. This device, with its capability of producing pure monochromatic light and a perfectly collimated light bundle, has had an immeasurable impact on the field of optics. Devices such as interferometers, laser range finders, holograms, laser disc players, and so many more, became an important part of optical engineering in the twentieth century.

2.7 Case History: Optics in the United States

Examination of the historical timeline in Fig. 2.5 shows that prior to the twentieth century there was little activity dealing with optics and optical engineering in the United States. Since the early part of the century, an increasing amount of research and engineering has been done in this country. Following the history of one typical optical company in the U.S. during that time period will provide some interesting insight into this period of growth and development.

In 1905 Frederick Kollmorgen left his position as optical designer with Ross, Ltd. of London to come to the U.S. to work with Keuffel & Esser in Hoboken, New Jersey, as a lens and instrument designer. Kollmorgen developed a keen interest in the subject of periscope design as it applies to military systems and the submarine. This led to several patents in this area being assigned to him in 1911. In 1914, with World War I eminent, the U.S. Navy contacted K&E in the hope that they might become the supplier of periscopes for the submarine fleet that was then under development. K&E declined the proposal, but Kollmorgen's interest in the project led him to pursue another approach. At the time, Kollmorgen had been dealing with a small optical company in New York City named Eastern Optical Co. He approached them with the proposition that he and they join forces to become the Navy's submarine periscope supplier. The favorable business aspects of such a venture were immediately obvious, and the

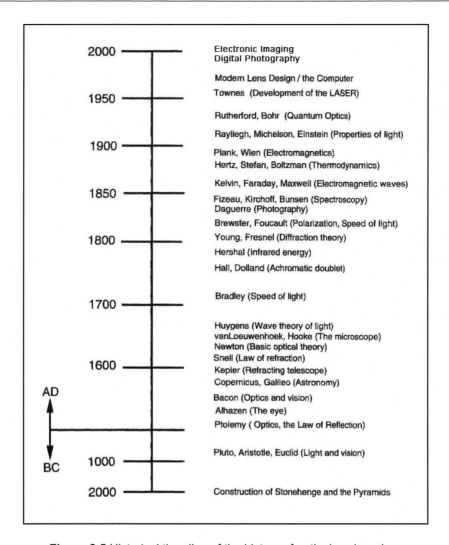

Figure 2.5 Historical time line of the history of optical engineering.

Kollmorgen Optical Company (KOC) was formed early in 1916. The wisdom of this decision is validated by the fact that every submarine periscope deployed by the U.S. Navy since that date has been designed by Kollmorgen personnel, and well over 95% of these periscopes have been manufactured by KOC. During World War I, the manufacture of periscopes and other military instruments resulted in an increase in the number of employees at KOC to nearly 100 and a move from Manhattan to a larger facility across the river in Brooklyn.

With the end of the war, KOC branched into the manufacture of commercial optical products. This included a very successful line of motion picture projection lenses. Late in the 1920s the great depression began and the very survival of KOC was threatened. In fact, for a brief period in 1932 the company was completely closed. In the mid-1930s the government sponsored several programs dealing with the development of military instruments, and KOC began to rebuild.

In the late 1930s they were involved in the design and manufacture of periscopes, drift meters, bomb sights, and navigational instruments. With the start of World War II in the early 1940s manufacturing activity grew at a frantic pace. Between 1940 and 1945, the number of KOC employees grew from 50 to more than 600. At the point of peak production in 1944, submarine periscopes were being shipped at a rate of one per day. In addition, hundreds of less-complex components and instruments were being produced daily. In 1945, the war ended and with it the demand for large quantities of military equipment. KOC again went through the "swords to plowshares" routine, again finding the motion picture projection lens to represent a valuable commercial product line. Other commercial products at the time included rifle scopes and monocular viewers. During the 1950s there was considerable activity involving the development of new motion picture formats such as CinemaScope and Cinerama. KOC was involved in much of this development activity, which would eventually lead to KOC receiving several Academy Awards for technical excellence in the field. During the 1960s the Cold War and the start of the space race led KOC into a number of interesting fields, most involving periscope designs of one form or another. They also developed a line of optical tooling instruments that were well received at the time. Meanwhile, the submarine periscope business continued to grow, spurred by the Polaris and subsequent fleet ballistic missile programs of the U.S. Navy. This periscope design work peaked in the late 1960s when KOC designed and produced a state-of-the-art submarine periscope designed specifically for photographic reconnaissance missions. This periscope served effectively for over 40 years. It was the design of this periscope in the 1960s that brought the designers of KOC into the modern age of computerized optical design.

Prior to the late 1930s, all lens design calculations at KOC had been done by hand, under the close scrutiny of the company's founder, Frederick Kollmorgen. During World War II, the design staff was expanded to include several additional optical engineers, along with a team of assistants working with mechanical calculators. In the 1950s the earliest of electronic computers came on the scene to enhance the capabilities of the design staff. It was not until the 1960s that the introduction of the larger mainframe computer, along with a modern lens design software package, would finally eliminate the need for a staff of calculating assistants to work with and support the optical engineers.

In recent years KOC has continued to grow and diversify, now a part of a larger multinational corporation. A key division of this corporation is directly descended from the KOC that was founded by Frederick Kollmorgen back in 1916. Here a line of sophisticated military systems and devices are designed and produced. This product line includes submarine periscopes for the U.S. and several allied navies, along with a variety of other electro-optical systems intended primarily for military applications.

2.8 The Hubble Space Telescope

Until recently, a declaration that one worked as an optical engineer would nearly always lead to a discussion involving the fit and function of eyeglasses. Early in the 1990s, thanks to the Hubble Space Telescope program, all of that changed abruptly. Suddenly the eyeglass discussion had been replaced by one dealing with "What happened to the Hubble?" and "Did you have anything to do with it?" A review of what has come to be known by some as the "Hubble Trouble" will serve well to illustrate the function of today's optical engineer and, unfortunately, several of the things that can go drastically wrong with a complex optical program.

Since the earliest times, the quality of images viewed by astronomers has been limited by the fact that the celestial objects had to be observed through the earth's atmosphere. Inconsistencies in the density of the atmosphere, combined with impurities and local stray light, tended to reduce the image quality of even the finest telescopes. This has resulted in the location of most major observatories away from city lights and on the highest available mountain tops. This concept would be epitomized by the placing of a large telescope in orbit around the earth, above its atmosphere. Following several preliminary efforts which realized varying degrees of success, the Hubble Space Telescope (HST) program finally got under way in 1977. A space shuttle would be used to deliver into orbit a Cassegrain telescope with an aperture of 2.4 m. After numerous delays, including the Challenger disaster in 1986, the HST was finally launched in April of 1990. In June it was found that the telescope was not performing as anticipated. The image quality was not border-line bad . . . it was *terrible!* It was concluded by examination of the images being produced that one of the two telescope mirrors was faulty in its basic shape. Examination of test data and procedures revealed that it was the primary mirror that was faulty because it had been manufactured using a test apparatus that was seriously in error.

The primary mirror of the HST is a concave hyperbola in cross section. In order to test for this shape as the mirror was being polished, a precision null-lens assembly was designed and produced that would generate a wavefront that would precisely match that hyperbolic shape. During the manufacture of that null-lens assembly, a fundamental mistake was made that resulted in an error in the axial location of one of the null-lens elements. The resulting null-lens assembly was not per the nominal design, and the optician charged with polishing the primary mirror would shape that surface such that it precisely matched the erroneous wavefront produced by that null lens. In hindsight, there are any number of steps that could (or should) have been taken that would have uncovered the problem. Indeed, several incidents reported to have occurred during manufacture were clear indicators that there was a problem. Reportedly, schedule and budget considerations were such as to override these suspicions. A second, backup, primary mirror was being manufactured simultaneously by a second source, using independent methods and test equipment. An exchange of the two mirrors would have revealed the problem. Finally, a test might have been performed following assembly of the primary and secondary mirrors to confirm their combined image quality. This was not done, again allegedly, for reasons of budget and schedule. The final result was a telescope placed into earth orbit that did not come close to meeting the performance levels that were anticipated. One very important lesson to be derived from all of this is that care must be exercised at all levels of design, manufacture, and test, in order to assure a successful outcome. In this case, a relatively obscure lens element was placed 1.3 mm away from its proper axial location—this, in a system with major dimensions that are many meters in size. Everything else was done correctly (technically, if not politically), and the mirror surface was polished to an accuracy of 0.00001 mm, but using the wrong gauge. As a result, the image of a distant star was more than five-times larger than had been expected. This magnitude of error essentially offsets all of the advantages that are realized by placing the telescope in orbit above the earth's atmosphere. It is typically the job of the optical engineer to assure that blunders of this type and magnitude do not occur. Most familiar with the Hubble story agree that, had those optical engineering people involved in the program been allowed to execute additional tests that seemed to be required, then the ultimate problem would have been avoided.

It is gratifying to report that, following a multiyear NASA repair program, the HST optical performance was corrected and the project has been a success beyond all original expectations, returning spectacular data and images to earth for nearly twenty years.

2.9 Review and Summary

This chapter represents just a small portion of all that has transpired over the years relating to the history of optical engineering. It is obvious that the field has been in existence in some form for a long time and that many new and exciting developments have been witnessed in recent years. Today we are in the midst of a technological explosion related to optics. Whether we find ourselves working actively in the field, preparing ourselves to become workers, or are merely persons living among and benefiting from the fruits of this science, there is much going on here to be understood and appreciated. With that in mind, let's get on with our study of optics and optical engineering.

References

1. D. J. Lovell, *Practical Optics*, unpublished manuscript, (1960s).
2. R. Kingslake, *A History of Photographic Lenses*, Academic Press, San Diego (1989).
3. R. Kingslake, "Some Impasses in Applied Optics," *J. Opt. Soc. Am.,* **64**(2), 123– (1974).
4. J. R. Meyer-Arendt, *Introduction to Classical and Modern Optics*, Prentice-Hall, Englewood Cliffs, New Jersey (1972).

Chapter 3
Basic Concepts of Light

3.1 Light: an Elusive Topic

In order to begin to understand the field of optical engineering and the function of the optical engineer, it is essential that we first develop some sense of what light is and how it behaves. It has been the experience of many (including myself) that this understanding need be neither comprehensive nor precisely correct to be valuable to the optical engineer in the execution of his day-to-day responsibilities. After all, as we have seen in the previous chapter, many of the great minds throughout history have made significant contributions to the science of optics while maintaining incomplete, or in some cases completely erroneous, theories regarding the nature and behavior of light.

3.2 Understanding Light

First, we can safely and correctly state that light is energy. We know that light is electromagnetic energy and that, in terms of wavelength, it represents a very small part of the broad electromagnetic spectrum (see Fig. 3.1). It will be helpful if we pursue our understanding of the term *wavelength* a bit further at this point. In one general and very useful definition, light is that portion of the electromagnetic spectrum that the human visual system (the eye) is capable of detecting. That is to say, light is the electromagnetic energy that we can see. Under normal daytime conditions, the eye has a maximum sensitivity to light with a wavelength of 0.56 micrometers (μm). Let us convert this number (0.56 μm) to a more meaningful and comprehensible form. A meter represents a length of about 40 inches, or a little more than one yard. A millimeter is 1/1000th of a meter, or about the thickness of 10 pages in this book. A micrometer (often referred to as a micron) is 1/1000 of a millimeter, and the peak wavelength for visible light is about one half of that. Therefore, one wavelength of visible light has a length approximately equal to 1/200 the thickness of a single page in this book.

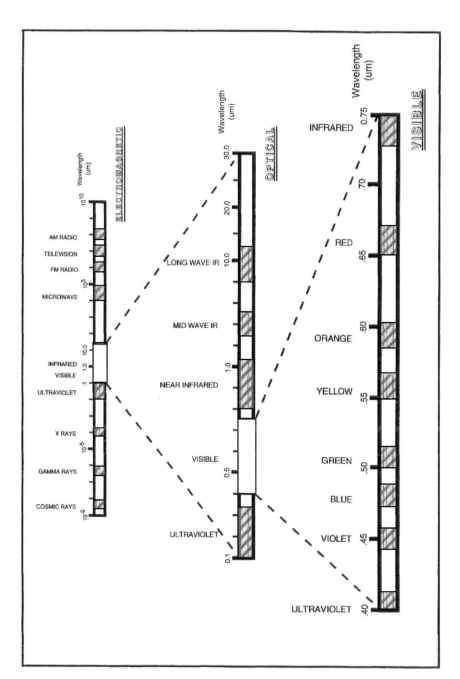

Figure 3.1 Wavelengths of the electromagnetic, optical, and visible spectra.

As we continue the study of optics we will find that the dimensions we encounter are most often represented by either very large or, as is the case for wavelength, very small numbers. In order to deal with these numbers it is helpful to understand and to use the system of scientific notation which incorporates the use of exponents. In its simplest form, this method presents us with a number between 1 and 10, followed by a notation indicating how many places the decimal point must be moved either to the left (−) or to the right (+). In this case, we might say that the peak visible wavelength is equal to 5.6×10^{-4} (0.00056) mm. For those readers who are not yet thinking and working in metric terms, this would convert to 2.2×10^{-5} (0.000022) in. As the first of many words of advice to the reader and would-be optical engineer, do make the conversion to thinking and working in the metric system. It will be most helpful and will enhance your effectiveness in nearly all areas of optical engineering.

As the wavelength of the energy (light) being collected by the eye increases or decreases, the color of that light as perceived by the eye will change. At the peak wavelength of 0.56 μm, the light will be seen as yellow. When the wavelength decreases to about 0.50 μm, the light appears to be green, while at 0.48 μm the color we see is blue. Moving in the other direction from the peak wavelength, at a wavelength of 0.60 μm, the light appears to be orange, and then, as the wavelength reaches 0.65 μm, we will see the light as red. This is what we refer to as the visible spectrum, ranging from violet (0.40 μm) to red (0.75 μm), with a peak sensitivity to the color yellow, at a wavelength of 0.56 μm. The visible spectrum, and its relationship to the electromagnetic spectrum, are shown in Fig. 3.1. Figure 3.2 shows the relative sensitivity of the eye as a function of wavelength, i.e., color.

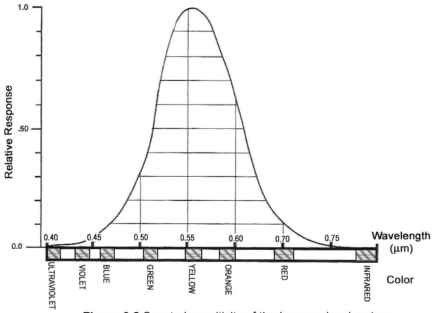

Figure 3.2 Spectral sensitivity of the human visual system.

3.3 Velocity, Wavelength, and Frequency

The wavelength concept may not be fully understood without some additional considerations. The topics of velocity and frequency and their relationship to wavelength must be introduced and discussed at this point. During our historical review of optics it was noted that, after many years of experimentation and several false starts, the velocity at which light travels in a vacuum was precisely determined to be 299,793 km/s. For purposes of this discussion, a rounded-off value of 300,000 km/s (3×10^8 m/s) will be used to represent the speed of light in air. The enormity of this speed is best visualized when we realize that, traveling at that speed, we would be able to travel around the world 7.5 times in a single second.

Another example that will be helpful in illustrating the speed involved deals with recent experiments conducted to measure the distance from Earth to the surface of the moon. After the astronauts had placed a mirror assembly on the surface of the moon, it was possible to project a pulse of laser energy from here on Earth to that mirror and to detect its reflection when it had returned to its point of origin. Knowing the time required for the round trip, and the speed at which the light traveled, it was then possible to calculate the distance from the laser to the mirror, i.e., Earth to moon. Quite incredibly, the time required for light to travel from the earth to the moon and return again was just 2.56 seconds! This data permits us to calculate the distance to the moon d as velocity v times time t, giving us a value of $d = (v)(t) = (300,000 \text{ km/s})(2.56/2 \text{ s}) = 384,000$ km.

In order to establish a relationship between the wavelength and the velocity of light that can be more easily understood and applied to our understanding of optics, the hypothetical example illustrated in Fig. 3.3 will be helpful. Consider a typical light source, such as the heated filament of an incandescent lamp. In order to further develop a sense of how light behaves, imagine this source to be pulsating, emitting light in a continuous stream within which the energy level is constantly and rapidly increasing to a maximum and then decreasing to a minimum. We have established that light travels away from the filament at a speed of 300,000 km/s. If a hypothetical energy level detector were placed as shown, at a fixed distance from the source, it would register those maximum and minimum energy levels as the wave impinged upon the detector. By definition, the wavelength of this energy is equal to the distance that the wave travels in the time that it takes the detector to record two consecutive maximum readings.

A heated filament is known to emit energy over a broad spectrum which includes all of the visible wavelengths and a part of the infrared band. To simplify this example, assume that a green filter has been placed between the source and the detector as shown in Fig. 3.3. Knowing the wavelength of the light passing through the green filter to be 0.50 μm (5×10^{-7} m/cycle), and the speed at which the light is traveling (3×10^8 m/s), we can compute the frequency at which the detector will register maximum readings using the following formula:

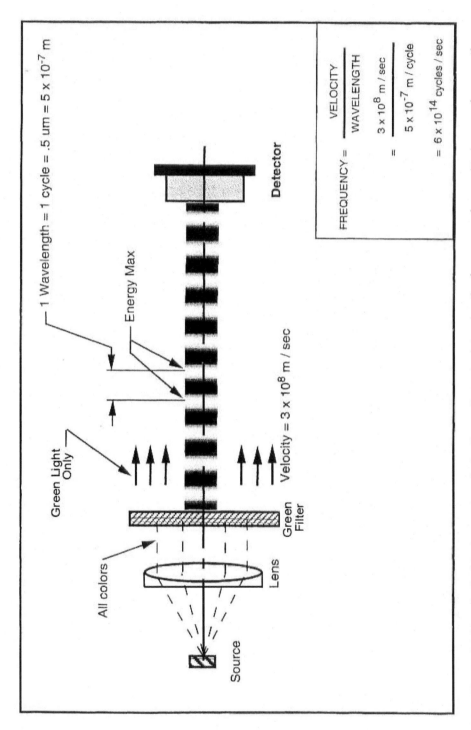

Figure 3.3 Hypothetical test procedure that will permit one to determine the frequency of an emitted wave of energy if the speed at which it travels and its wavelength are known.

$$\text{Frequency} = \text{velocity/wavelength}$$
$$\text{Frequency} = (3 \times 10^8 \text{ m/s})/(5 \times 10^{-7} \text{ m/cycle})$$
$$\text{Frequency} = 6 \times 10^{14} \text{ cycles/s (Hz)}.$$

In keeping with our goal, which is to develop a sense for what light is and how it behaves, we now can state that when we observe a green light source, the electromagnetic energy from that source is approaching our eyes at a velocity of 3×10^8 m/s, and its energy level is pulsating at a frequency of 6×10^{14} times each second.

As noted earlier, this discussion has neglected the difference in the velocity of light in air as compared to that in a vacuum. The fact is that light in air is slowed by about 0.03%, which is indeed negligible for these purposes. A second point along these same lines is that measurements have proven the velocity of electromagnetic energy in a vacuum is constant, regardless of its wavelength. In air the change in velocity with wavelength is so small as to be negligible for our purposes. Therefore, all the colors that we see travel at the same speed. It follows then that because frequency is inversely proportional to wavelength, the frequency for each color must be different. In the case of the visual spectrum, that frequency will range from 7.5×10^{14} Hz for blue light, to 6.0×10^{14} Hz for green light, down to 4.5×10^{14} Hz for red light.

Another example, taken from today's technology, will be helpful in establishing our sense for what light is and how it behaves. A common electro-optical instrument is the laser range finder (LRF). The LRF contains a very pure, monochromatic light source in the form of a laser, a set of projection optics including a shutter, a set of receiving optics, a detector, and a precision timing mechanism. The LRF functions by generating a pulse of laser energy that is projected onto a target where it is reflected back toward the LRF. That reflected light is collected by the receiving optics and imaged onto the detector. The precision timer measures the time required for the pulse to travel to the target and to return to the LRF. Knowing the velocity of light, it is then a simple matter to compute the distance to the target with great precision. One common LRF configuration uses a laser with a wavelength of 1.06 μm and a projected pulse duration of 20 ns. From this information we can generate a realistic description of that pulse, including its physical size and characteristics. To generate the pulse, the LRF shutter must be opened and closed in a precise manner, with a total elapsed open time of 20×10^{-9} s. We know that light travels at a speed of 3×10^8 m/s. Figure 3.4 illustrates the system being discussed. When the shutter has been open for the specified 20×10^{-9} s, the leading edge of that pulse will have traveled: $(20 \times 10^{-9}$ s$)$ $(3 \times 10^8$ m/s$) = 6$ m. The shutter will then close, and the result will be a 6-m-long pulse of laser energy, traveling toward the target at a speed of 3×10^8 m/s.

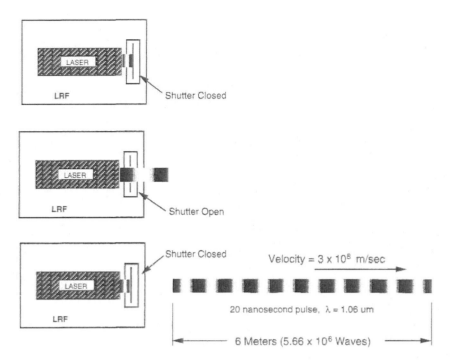

Figure 3.4 A laser range finder generates a short pulse of energy which is used to determine the distance to a target. Shown above is a schematic representation of that pulse.

We can determine one other significant characteristic of this pulse of laser energy. We have said that the energy emitted from the laser has a wavelength of 1.06 μm, or 1.06×10^{-6} m. It follows then that the 6-m-long pulse would contain:

$$(6.0 \text{ m/pulse})/(1.06 \times 10^{-6} \text{ m/wave}) = 5.66 \times 10^{6} \text{ waves}.$$

Returning to Fig. 3.4, we can now visualize and describe the 20-ns pulse from the LRF as a beam of energy with a wavelength of 1.06 μm and a total length of 6.0 m. That pulse contains 5.66×10^{6} cycles (waves) of this laser energy, and the entire pulse is traveling toward the target at a speed of 3×10^{8} m/s.

Exercises such as this are valuable in the sense that they serve to convert abstract concepts into real-world situations that we can more easily visualize and understand. With this understanding, the optical engineer will be better equipped to generate designs that effectively implement these concepts.

3.4 Wavefronts and Light Rays

In considering the path that light follows when it leaves a source, it will be simpler if we assume that light source to be quite small—what we will refer to as a point source. The light originating at a point source spreads out from that point, forming an expanding spherical wavefront. In the case of a point source at a great distance, such as a star, the radius of that spherical wavefront as it is detected here on earth will be infinite, i.e., the incident wavefront will be flat (plano).

In most optical engineering examples, the light from a source is described and dealt with in terms of light rays rather than wavefronts. Light rays are straight lines that originate at the source and, by definition, are perpendicular to the surface of the spherical wavefront (see Fig. 3.5).

3.5 Light Sources

A discussion of light sources would seem to involve an endless number of things, but, in reality, it is limited to just a few. The sun is obviously our most common source of light here on earth. Other common sources consist primarily of burning fuels and heated filaments. Other less-common sources would include fluorescent lamps, neon lights, and lasers. We are able to read the printed matter on these pages, not because they are illuminated from within, but because some external source has illuminated them with light that has then been reflected to our eyes. Just like other objects that we view or otherwise optically record during daylight hours, these objects are illuminated by the sun and then reflect a portion of that sunlight to our eyes. In viewing an extended scene, we detect varying levels of object brightness as a function of that objcct's reflectivity at the point being viewed. Likewise, we detect different colors when the object reflects different parts of the visible spectrum in differing amounts. The grass and leaves reflect primarily green light, and thus, they appear to be green. Similarly, the red convertible we admire has been painted with a product that reflects the red portion of the spectrum while absorbing all other colors.

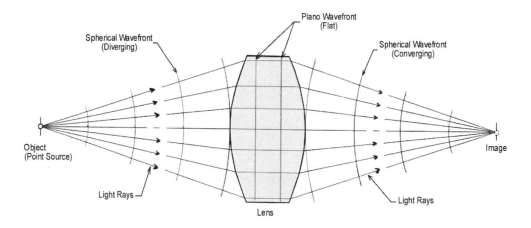

Figure 3.5 As light travels from a point source, it creates a spherical wavefront. For purposes of analysis, it is convenient to assume the existence of light rays. These are radial lines, originating at the source and traveling in a straight line, always perpendicular to the surface of the spherical wavefront.

For most basic purposes of optical engineering and analysis, it is valid to treat an object that is reflecting light from a source as a source itself. In this way we can handle subsequent analysis as if the light originated at that object and then traveled through the optical system and on to the detector for viewing or recording. It is important when performing such analysis to keep in mind that the apparent spectral characteristics of the object being viewed will change if the spectral output of the original source is changed. For example, if a white box is viewed in sunlight, it appears to be white. If viewed at night when illuminated by a sodium arc lamp whose spectral output is primarily orange, then the white box will appear to be orange.

In a vast majority of cases the object under consideration is being illuminated by one of three common light sources, or *illuminants*. They are: a tungsten filament, direct sunlight, or average daylight. These sources have been designated "standard illuminants A, B, and C" respectively, by the International Commission of Illumination, for purposes of standardized colorimetry discussions. Figure 3.6 shows the relative spectral energy output of these three standard sources. These data will often be found useful in determining the spectral nature of an object during the analysis of an optical system.

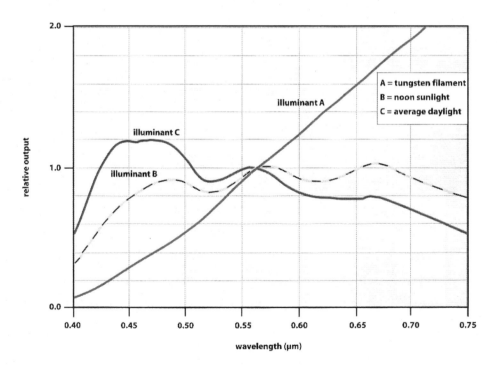

Figure 3.6 For purposes of a standardized discussion of colorimetry, CIE has established three standard sources of illumination. They are: tungsten filament, noon sunlight, and average daylight. These are referred to as standard illuminants A, B, and C, respectively. The spectral energy distribution of these three standard illuminants is shown here.

3.6 Behavior of Light Rays

A light ray is not so much a thing as it is a concept. Earlier we defined a light ray as a straight line, originating at a point on the object and extending to a point on the wavefront that has been generated by that object. The light ray concept is particularly useful in that it becomes much easier for us to visualize and calculate the behavior of light as it travels from an object through an optical system, and then forms the final image, usually at a detector. A light source does not emit light rays; rather, it emits a spherical wavefront that can be conveniently represented and illustrated using light rays. This light-ray concept was illustrated earlier in Fig. 3.5.

The value of dealing with light rays as opposed to waves and wavefronts will first be demonstrated by a discussion of reflectors (mirrors) and how they modify the light that is incident upon them. The simplest case would be a flat, or plano, mirror. In Fig. 3.7, there is a point source of light labeled *object* and, at a distance *S* to the right, there is a flat mirror. The object may be an original source, such as a lamp, or it may be (and more commonly is) a point on an extended object that is reflecting light from some original source. A second point worth noting is that, while the object may be emitting light into any portion of a complete sphere, we are only concerned with that portion of the emitted light that intersects the optics being analyzed, in this case the flat mirror.

The purpose of this example is to generate a graphic representation of how the light from an object is affected by the mirror, and to allow us to develop an understanding of that behavior with which we are comfortable. The following basic law of optics applies:

The Law of Reflection:
> When an incident ray of light is reflected from a surface, that incident ray, the normal to that surface, and that reflected ray, will all fall in the same plane, and the angle of incidence will be equal to the angle of reflection.

In Fig. 3.7, the plane containing the incident and reflected light rays is chosen to be the plane of the paper; thus, the angle of incidence *i* will be equal to the angle of reflection *r* as shown. Geometric construction (or doing the math) will lead to the same conclusion, namely, that an image of the object will be formed on the optical axis (an extension of that light ray that travels from the object, normal to the mirror surface), at a distance *S'* to the right of the mirror surface. In the case of a flat mirror, *S'* will be equal to *S*.

This very basic exercise illustrates the value of the light-ray concept and how easily it can be applied to the analysis of a simple ray-trace problem. By tracing just a single ray from the object, it is possible to determine the exact location of the image that will be formed by the mirror.

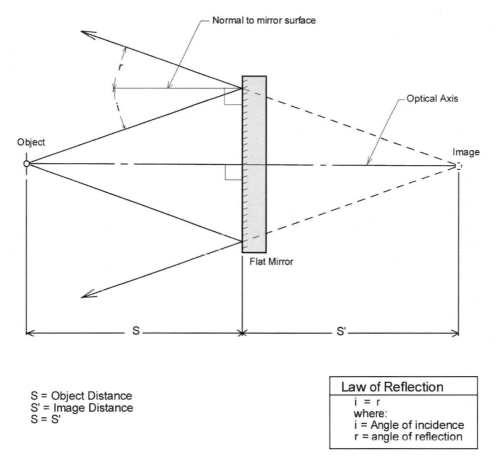

S = Object Distance
S' = Image Distance
S = S'

Law of Reflection
i = r
where:
i = Angle of incidence
r = angle of reflection

Figure 3.7 This figure shows the use of the light-ray concept to analyze the interaction between light from a point source object and a plano (flat) mirror.

Optical engineering often involves establishing a basic condition, such as the flat mirror, and then modifying certain factors and determining a new result. For example, we might be interested in the effect should the distance S from the object to the flat mirror be modified. In this case, each change to S would produce an equal change to S'. Consider a real world example. When looking into a mirror, assume there is some detail on our face that we wish to examine more closely. We would move closer to the mirror, thus reducing the value of S and S'. The result is that for each inch we move closer to the mirror, the image we wish to examine is moved two inches closer to our eye.

A second variation of the simple mirror example—and one that is much more significant and interesting—would be the shape of the mirror surface. Typically, a mirror will be either flat, concave, or convex. When the mirror surface is assumed to be convex, as in Fig. 3.8, it can be seen that the first obvious result is that the normal to the mirror surface, where the ray is incident, is now tilted upward rather than being parallel to the optical axis as it was for the flat mirror.

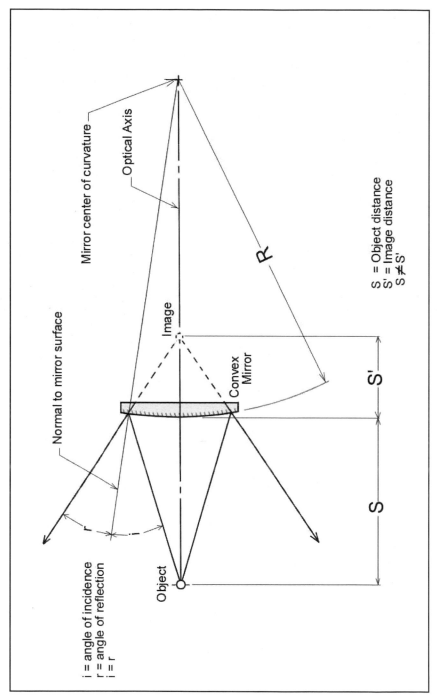

Figure 3.8 This figure shows the use of the light-ray concept to analyze the interaction between light from a point source object and a curved (convex) mirror.

Note that when extended, the normal will intersect the optical axis at the center of curvature of the mirror surface. Since the law of reflection will apply, the reflected ray now leaves the mirror surface at a steeper angle than was the case for the flat mirror. The result is that the image is formed closer to the mirror, without the object having been moved at all. Here again, the tracing of a single ray allows the determination of image location.

This convex mirror example explains, in part, the origin of that familiar phrase "Objects in mirror are closer than they appear," which we have all seen etched on the passenger side (convex) mirror on our passenger car. I say "in part" because if it were purely a matter of the image having been moved closer to the mirror, then the objects would actually be *farther* than they appear. However, what has happened simultaneously is that the convex mirror has reduced the size of the image by a greater factor than it has moved that image closer to the viewer. The result is that while the image is indeed closer, it is also considerably smaller than it would be were the mirror surface flat. When we see an image of a familiar object (such as a car in the mirror), we judge its distance based on the size of that image. If the image appears smaller (than it would reflected in a flat mirror) we assume that the object is at a greater distance . . . but in this case it is not—it is closer than it appears. This example is not as complex as it might seem. Try to accept it as interesting rather than confusing. We will be spending much more time on the subject of object-image relationships and their determination in future chapters. Again, the key objective at this point is to develop a certain level of comfort when dealing with light rays as they pass from an object, reflect from a mirror, and then form an image.

In Fig. 3.9 we again see the same point source object as in earlier examples, with a concave mirror at the same distance S. Again, the normal to the mirror surface at the point where the light ray strikes is represented by a line from that point to the center of curvature of the concave mirror surface. While this normal was parallel to the optical axis for the flat mirror, it is now tilted downward toward the optical axis. As a result, we see that the reflected ray will also be tilted downward, such that the image is formed to the left of the object at a distance S' from the mirror. Again, the law of reflection has been applied to the tracing of a single ray, allowing us to determine the exact image location.

While the example involving a concave mirror as shown in Fig. 3.9 is being considered, it will be useful to discuss a special case. It will be found that as we move the object closer to the mirror by a small amount, the location of the image will move away from the mirror by a considerably larger amount. In the extreme case, when the object is moved to the midpoint between the mirror and its center of curvature, then the reflected ray will be essentially parallel to the optical axis and the image will be formed (theoretically) at an infinite distance from the mirror. In this specific case the source is located at the focal point of the concave mirror and the reflected light is said to be *collimated*, i.e., projected to infinity. Again, this condition and these terms will be covered in greater depth as they are encountered in future chapters. At this point we are primarily concerned with the

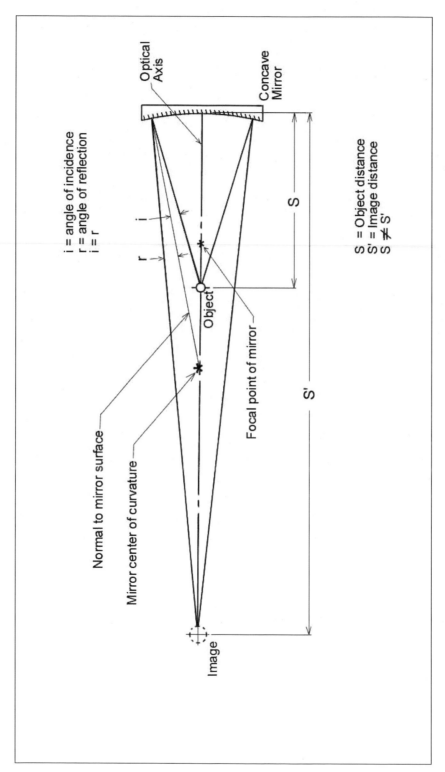

Figure 3.9 This figure shows the use of the light ray concept to analyze the interaction between light from a point source object and a concave mirror.

light ray concept and how its application has conveniently allowed us to visualize and understand the interaction between an object and a basic optical component . . . a mirror.

3.7 Refraction

Starting again with a basic definition, refraction is "the deflection of a propagating light wave." The relationship between light waves and light rays has already been introduced. It will be helpful now to establish a sense for the relationship between a wavefront and its associated light rays as refraction takes place. In order to do this, it will be helpful to first understand the fundamental concept of the *index of refraction*. We have stated that the speed of light in air is about 3×10^8 m/s. As light travels through a medium other than air its speed is reduced. The index of refraction of a material is determined by measuring the speed of light in that material and dividing that speed into the speed of light in air. Two very common examples of refraction encountered in optics discussions involve light traveling through water and light traveling through glass. Typical measurements for water yield a speed of light equal to 2.25×10^8 m/s. The index of refraction n for water is then found to be:

$$n_{\text{water}} = \left(3 \times 10^8 \text{ m/s}\right) / \left(2.25 \times 10^8 \text{ m/s}\right) = 1.333.$$

In the case of optical glass, there are a number of different glass types, each with a different index of refraction. The most common optical glass type encountered (Crown glass) has an index of refraction of 1.52. From this value, we may conclude that the speed of light traveling through this typical optical glass will be 1.97×10^8 m/s. In other words, light leaving a source in air will be traveling at a speed of 3×10^8 m/s. When that light enters a typical optical element with an index of 1.52, the speed at which it is traveling will be reduced to 1.97×10^8 m/s. Figure 3.10 graphically shows the relationship between the speed of light in a vacuum, where the index of refraction is 1.00, and the speed of light in a variety of other materials where the index is greater than 1.00.

With the relationship between the speed of light and the index of refraction as background, let's now consider the refraction (deflection) of a spherical wavefront when it enters a glass block of a higher index material as is illustrated in Fig. 3.11. The illustration shows a point source of light and the spherical wavefront that is traveling from that source in a left to right direction. When the spherical wavefront encounters the glass block, it will be noted that the point on the wavefront that is on the optical axis will be the first to enter the glass. At this point, that portion of the wavefront is slowed, while that portion remaining outside of the glass is not. The result is that when the entire wavefront is inside the glass block, its curvature will have been reduced, or flattened. The radius of the wavefront within the glass block is now greater than it would have been if the glass were not present. In other words, the wavefront has been deflected, or refracted.

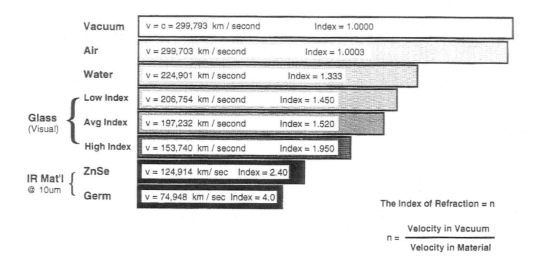

Figure 3.10 The speed at which light travels is a function of the material in which it is traveling. Shown above is the speed and resulting index of refraction for a number of common optical materials.

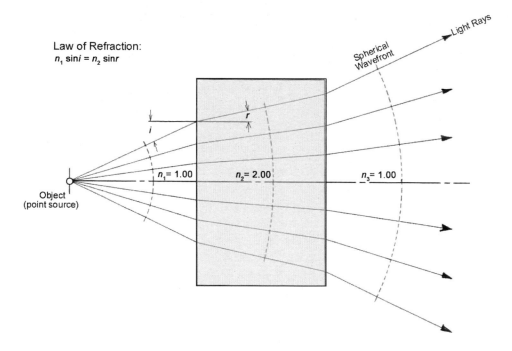

Figure 3.11 As light travels from a point source it creates a spherical wavefront. When that wavefront enters a second medium, its shape is changed. The behavior of associated light rays is defined by the law of refraction, also known as "Snell's law."

Likewise, the associated light rays seen in the figure will bend or refract as they enter the glass block. The amount of this bending may be precisely calculated using the law of refraction, or Snell's law, which states the following:

$$n_1 \sin i = n_2 \sin r,$$

where:
n_1 = index of refraction to the left of the boundary,
n_2 = index of refraction to the right of the boundary,
$\sin i$ = sine of the angle of incidence, and
$\sin r$ = sine of the angle of refraction.

Also, the normal to the boundary, the incident ray, and the refracted ray all lie in a common plane.

It is interesting to note that, for the case illustrated, the glass block is shown with its entrance and exit faces flat and parallel. As a result, the amount of refraction for the emerging rays will be equal to the amount at entering, thus the angle of the ray relative to the optical axis will be restored to its original value. This sense of how refraction occurs when a wavefront of light passes from one medium to the next will be helpful in the appreciation of numerous optical phenomenon that are to be covered in upcoming chapters.

3.8 Refraction by a Lens

The case illustrated in Fig. 3.12(a) is similar to that shown in Fig. 3.11, except the point source of light has been assumed to be located at a great distance to the left, perhaps a star. As a result, the incident wavefront is flat (radius = infinity) and the light rays from that source are parallel to the optical axis. These rays strike the glass surface with an angle of incidence equal to zero and, as a result, they are not refracted (bent) as they enter and leave the glass block. This case illustrates a lens with zero optical power, i.e., a simple window.

In Fig. 3.12(b) we have the case where the first surface of the glass block has been modified to have a convex shape. As the incident wavefront reaches this glass block (now a lens), it can be seen that the central portion of that wavefront will be slowed while the outer portion is not affected. The result is the conversion of the wavefront from flat to curved as it enters the glass. Considering the behavior of the light rays, it can be seen that those light rays closer to the optical axis have a very small angle of incidence and therefore very little refraction (bending) of those rays takes place. As the distance of the ray from the optical axis increases, a proportionately larger amount of refraction takes place. The result is that the bundle of parallel rays is converted into a bundle of converging rays, indicative of the curved wavefront traveling through the lens. Even though the second surface of the lens is flat, it can be seen that the emerging wavefront gains additional convergence as it leaves the second surface. This case illustrates the behavior of a classic plano-convex positive lens element.

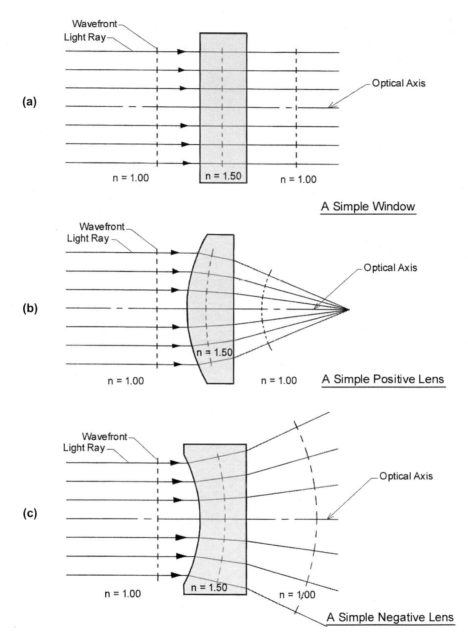

Figure 3.12 As light travels from a distant point source, it arrives as a plano (flat) wavefront. When that wavefront enters an optical component, its shape is modified. The amount and nature of that modification depends on the characteristics of the optical component.

Finally, in the example illustrated in Fig. 3.12(c), the glass block has been configured so that its entrance face is now concave. In this case, in a manner similar to the convex entrance face, the incident plano wavefront is again refracted such that it becomes curved (now diverging) as it travels through the lens. As the wavefront emerges from the flat second surface of the lens, it again becomes more steeply curved. In this case the wavefront and its associated light rays are diverging, illustrating the behavior of a classic plano-concave negative lens element.

This description of refraction and these simple lens forms will be expanded upon in subsequent chapters. At the risk of redundancy, I again remind the reader that the objective at this point has been to present these basic concepts in such a way that the reader will gain a comfortable feel for how light behaves when it encounters a simple reflecting or refracting optical element.

3.9 Dispersion and Color

In the case of light rays, dispersion can be defined as the separation of light into its color components by refraction. When refraction takes place at the boundary between two optical materials, it will nearly always be accompanied by some amount of dispersion. It will be useful to understand the reasons for this and some of the situations that will result. Figure 3.13 is similar to the window

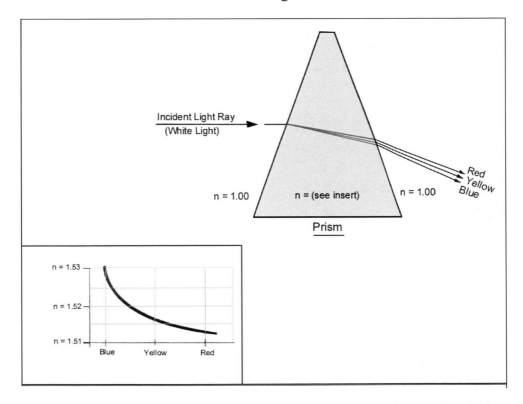

Figure 3.13 A light ray that is incident on a prism will be refracted as it enters. In addition, because the index of refraction of the glass varies with color, the amount of refraction will also vary with color. This dispersion of the incident white light will also occur as the light emerges from the second surface of the prism.

example that was shown and discussed earlier, except in this case the entrance and exit faces of the glass block are not parallel to each other. A block of optical glass with this configuration is referred to as a prism.

In Fig. 3.13, a single ray of white light (containing all colors) is shown striking the entrance face of the prism. Application of the law of refraction permits us to conclude that the ray will be refracted (bent) as it enters the prism. This seems straightforward, but there is a complication. When the index of refraction for an optical glass is measured precisely, it is found to vary slightly depending on the wavelength (color) of the light being measured. The inset in Fig. 3.13 shows an index versus wavelength curve for a typical optical glass. From this curve it can be seen that the index of refraction *n* is somewhat greater in the blue region of the spectrum while it is less toward the red end. Since the index in the incident medium (air) is essentially constant at 1.00 for all colors, and the index of the glass changes with color, it follows that by application of the law of refraction the angle of refraction will also change as a function of the wavelength of the incident light. In this example, if we assume that the incident light ray contains all colors of the visible spectrum, then the refracted light within the prism will be spread over a finite range of angles as shown. This spreading of light into its separate spectral components is known as *dispersion*. When the light rays emerge from the prism they are subject to additional refraction and dispersion. This phenomenon is frequently used in an optical instrument where separation of white light into its many spectral components is desired. More often, dispersion is found to be a defect in an optical system that leads to degraded performance. In that case it becomes the task of the optical designer to configure the system such that the amount of residual dispersion, i.e., chromatic aberration, is acceptable in terms of final system performance.

3.10 Diffraction of Light

In this discussion of the characteristics and behavior of light, the subjects covered to this point have been relatively easy to visualize and to understand. A light ray can be imagined to bounce off the surface of a mirror much as a billiard ball bounces off the cushion of a billiards table. The refraction and dispersion of a light ray can be directly related to the speed at which light travels and the change that occurs as the light moves from the air into a block of glass. In discussing diffraction, the theory and behavior of diffracted light is more difficult to visualize, but no less important to our study of optics.

In an optical system, an aperture is any device that limits the size and shape (usually round) of the light bundle that will be passed. When a wavefront passes through an aperture, the light at the edge of that aperture is diffracted. Diffraction is a phenomenon similar to dispersion, but it is not related to variations in the wavelength of the light. Because it is derived from the wave theory of light, a ray-trace analogy of diffraction is not a reasonable thing to attempt. Rather than delving into exactly what is happening to this light at the edge of the aperture, it will be more productive to deal with diffraction in terms of the resulting impact that it has on image quality of the optical system. The most common manifestation of diffraction will be seen when evaluating the optical performance of an optical system that is very nearly perfect. The lens shown in Fig. 3.14 will be assumed to be essentially perfect in terms of its image quality. Ray-trace analysis of this lens would indicate that all rays within a bundle parallel to the optical axis will be refracted by the lens such that they all intersect precisely at a common focal point. This ray analysis, also known as geometric analysis, would indicate that if a microscope were used to view the image formed by this lens of a distant point source, such as a star, a very bright point image of essentially zero diameter would be seen.

In reality, for a perfect lens such as this, the resulting image will be a bright central spot of finite size, surrounded by a series of concentric rings of rapidly decreasing brightness. The image just described is referred to as the classic *Airy disk*, which is the result of diffraction which occurs at the circular aperture of the lens. The exact size and makeup of the Airy disk can be determined using the data presented in Fig. 3.14.

This basic concept of diffraction by a lens aperture is very important because it is unavoidable. There is no way to produce an optical system that will perform better than the limits that are predicted by diffraction. This is really not a bad thing, because it gives the designer and the lens manufacturer a realistic target to strive for. Once the optical design has been optimized by the lens designer (and manufacturing tolerances have been assigned) such that all light rays traced fall within the diameter of the central spot of the Airy disk, there is little to be gained by further improvement of the design or reduction of manufacturing tolerances.

A second diffraction effect that is frequently encountered in optical system design comes about when the lens aperture contains an obscuration. This will usually occur in the form of an opaque circular blockage that is centered on the lens aperture (see Fig. 3.15). The most common occurrence of this is in a two-mirror telescope objective, such as a Cassagrain mirror system. Again, the presence of such an obscuration only becomes important when the basic lens design is otherwise nearly perfect. In such a case the lens would again produce a diffraction pattern in the form of an Airy disk. However, the presence of an obscuration will cause some of the energy within the central bright spot to be shifted out into the concentric rings. When the ratio of the obscuration diameter to that of the aperture is between one-third and half, the resulting degradation of image quality will be significant and must be taken into account when generating system performance specifications. The table shown in Fig. 3.15 indicates the magnitude of the energy shift. The result, in terms of image quality, will be a reduction in final image quality and contrast. For these reasons, systems with obscurations greater than half are rarely used.

3.11 Review and Summary

It would be presumptuous to say that this chapter represents all that we need to know about light, its characteristics, and its behavior. Suffice it to say that, if you have reached that point where you are comfortable with all of the concepts

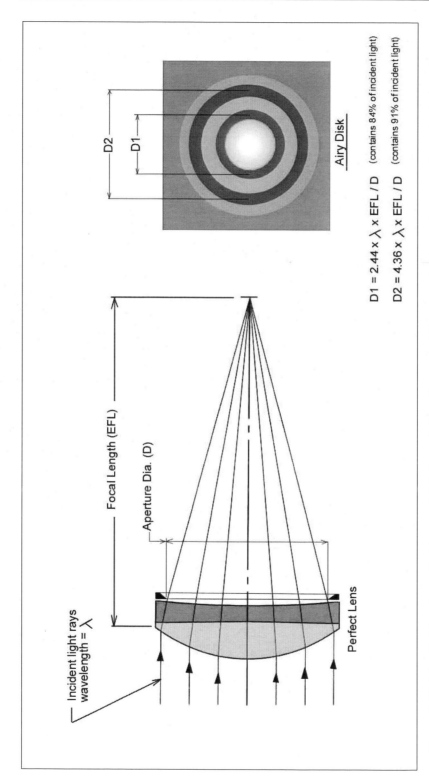

Figure 3.14 A "perfect lens," one that is completely free of aberrations, would be expected to produce a perfect point image of a distant point source, such as a star. In reality, diffraction effects at the lens aperture lead to the creation of the Airy disk rather than a perfect point image.

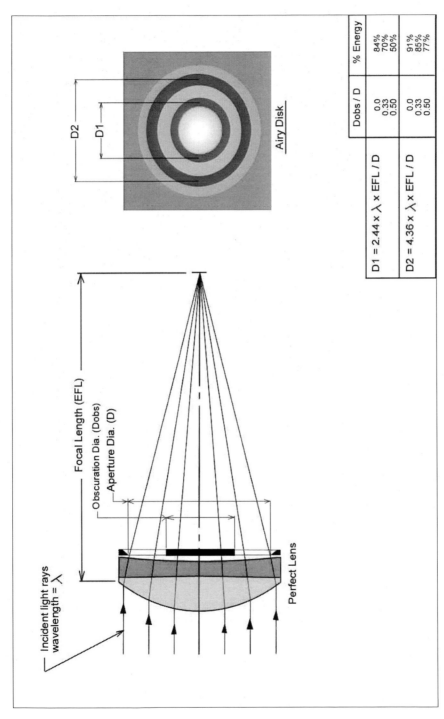

	Dobs / D	% Energy
D1 = 2.44 x λ x EFL / D	0.0 0.33 0.50	84% 70% 50%
D2 = 4.36 x λ x EFL / D	0.0 0.33 0.50	91% 85% 77%

Figure 3.15 Frequently an optical system will have a circular aperture with an opaque obscuration at its center. In that case the diffraction effects will result in a reduction of the energy level within the central spot of the Airy disk, with that energy shifted outward to the concentric rings. The result will be a significant drop in the contrast of the final image.

presented here, and you have developed a good feeling for what light is and how it behaves, then you are well equipped to deal with the remainder of this book and many (indeed most) of the optical engineering problems that you may encounter in the future. Let's now review the key points that have been covered in this chapter with regard to light.

First, light is energy. It represents a very small portion of the electromagnetic spectrum, which ranges from cosmic rays with a wavelength of 10^{-9} μm, to radio waves with a wavelength of 10^{10} μm. within this electromagnetic spectrum we find the optical spectrum, which contains energy from the ultraviolet to the infrared, representing wavelengths from 0.1 μm to 30 μm, respectively. Finally, and of greatest interest to most of us, within the optical spectrum we find the visible spectrum. This contains energy that we can see, ranging in color from deep violet to deep red, with a corresponding wavelength range from 0.40 μm to 0.75 μm. In the center of this visible spectrum is the color yellow with a wavelength of 0.56 μm. This is the peak sensitivity point of the human visual system.

Light is energy that travels very fast, so fast that its speed was thought by many, for many years, to be infinite. Over the years, methods have been developed to measure its exact speed. Best measurements to date indicate that speed to be 3×10^8 m/s, or 186,000 miles/s.

When light is emitted from a point source, it can be thought of as traveling in the form of a pulsating, rapidly increasing spherical wavefront. The frequency at which the energy level pulsates is a function of the wavelength of that energy. If the point source is located at a great distance from the optics being considered, it is valid to assume the incident wavefront to be flat (plano). The light-ray concept facilitates our understanding and analysis of wavefront behavior. A light ray is (by definition) a straight line, originating at the source and always perpendicular to the wavefront surface at its point of intersection.

When a light ray encounters an optical component, there are several possible reactions. On striking a mirror, the light will be reflected from the mirror surface. The shape of the mirror surface (flat, convex, concave), and the law of reflection will determine the direction taken by the reflected light ray. When a light ray is incident on a transparent block of glass, that light ray will enter the glass. The speed at which light travels within the glass is indicated by the index of refraction of the glass. The law of refraction permits us to precisely calculate the new direction that will be taken by the light ray after it enters the glass. A curved surface on the entrance surface of the glass block will result in a change in wavefront curvature; the resulting optical component is a lens.

Optical glass is found to have an index of refraction that varies with wavelength. As a result, when a light ray is refracted at an air-to-glass boundary, the amount of refraction will be a function of wavelength. Thus, if an incident light ray is made up of white light (light containing all colors), then the phenomenon of dispersion will split that ray into its many component colors.

Finally, we have shown in this chapter that when a wavefront passes through a limiting aperture, some of that energy will be diffracted, i.e., deviated, from the path that would have been predicted by conventional ray-trace analysis. This diffraction is present in all optical systems . . . when all other limitations have been overcome, the image quality of a perfect optical system will be limited by these diffraction effects.

From among the many facts and figures presented in this chapter, the following are frequently referred to and, as a result, are worth remembering:

Speed of light: 3×10^8 m/s (186,000 miles/s)
Visible spectrum: 0.40 μm (blue), 0.56 μm (yellow), 0.75 μm (red)
Law of reflection: $i = r$
Law of refraction: $n_1 \sin i = n_2 \sin r$
Index of refraction: $V_{vac}/V_{med} = 1.33$ (water), 1.52 (glass)
Airy disk diameter: $2.44 \times \lambda \times (f/\#)$ (first dark ring).

Chapter 4
Thin-Lens Theory

4.1 Definition of a Thin Lens

A *thin lens* is a theoretical lens whose thickness is assumed to be zero and, therefore, negligible during thin-lens calculations. The thin lens is a design tool used to simulate a real lens during the preliminary stages of optical system design and analysis. The concept is particularly valuable because it enables the optical engineer to quickly establish many basic characteristics of a lens or optical system design. By assuming a lens form where the thickness is zero, the formulas used to calculate object and image relationships are greatly simplified. The drawback to the thin-lens approach is that it is not possible to determine image quality without including actual lens thickness (and other lens data) in the calculations. As a result, while it is possible to establish many valuable facts about an optical system by the application of thin-lens theory and formulas, the ultimate quality of the final image can, at best, only be estimated.

4.2 Properties of a Thin Lens

Figure 4.1 is an illustration of a positive thin lens. Any lens system analysis must start with several known factors that will generally be provided by the end user, i.e., the customer. From these given starting factors it will be possible, using thin-lens theory and formulas, to generate the missing information required to completely describe the final lens system. In the case shown in Fig. 4.1, for example, it is given that the system detector (image size) will be 25 mm in diameter and that the full field of view of the lens will be 10 deg. Other system considerations dealing with detector sensitivity—thus required image brightness—dictate that a lens speed (*f*/#) of *f*/2.0 will be required. Since we know the image size and the field of view, applying the formulas shown in Fig. 4.1, we can calculate the effective focal length (EFL) of the lens. The image size dimension used for this calculation is measured from the optical axis and is designated as y_1. In this case, the maximum value for y_1 is 12.5 mm. Since the half field of view θ is 5 deg, the EFL of the thin lens can be determined using the formula:

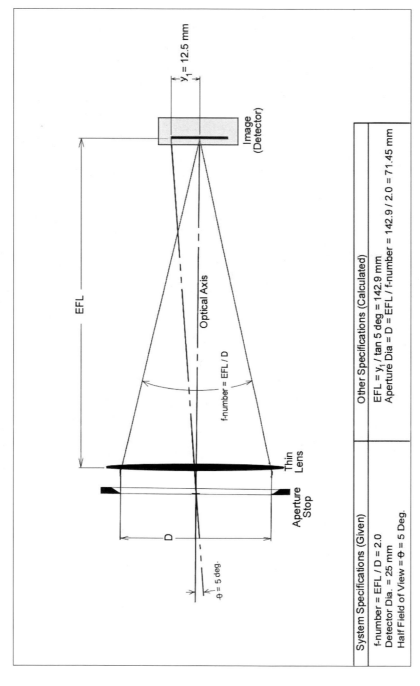

Figure 4.1 Shown here are the basic parameters of a thin lens. In this example, three given specifications allow calculation of two others.

$$\text{EFL} = \frac{y_1}{\tan\theta} = \frac{12.5\,\text{mm}}{\tan(5\,\text{deg})} = 142.9\,\text{mm}.$$

The diameter of the thin lens is a function of its EFL and $f/\#$ and is derived as follows:

$$\text{Diameter} = \frac{\text{EFL}}{f/\#} = \frac{142.9\,\text{mm}}{2.0} = 71.45\,\text{mm}.$$

Having now established a complete set of thin-lens specifications for the lens system, our next step would be to select a commercially available lens that meets these specifications and to compare its image quality with the requirements of the system (normally established by the image quality limitations of the detector). If, for any reason, the commercially available lens is not acceptable, then a custom lens design will be in order.

4.3 Aperture Stop, Entrance and Exit Pupils, and Field Stop

Any lens assembly, regardless of its complexity, will have one lens aperture or a mechanical component whose size will limit the diameter of the axial light bundle (that bundle of light rays originating at the center of the object) that is allowed to pass through the optics and onto the image surface. For example, in the thin lens shown in Fig. 4.1, a mechanical opening labeled *aperture stop* is shown in front of the lens. It can be seen that, without this stop in place, the lens would be able to accept and pass a axial-light-bundle diameter larger than *D*. Frequently the aperture stop will take the form of an adjustable iris diaphragm. This allows the effective diameter, thus the $f/\#$, of the lens to be varied such that the brightness of the image formed at the detector is constant, regardless of the brightness of the scene being imaged.

The entrance and exit pupils of a lens system are directly related to the aperture stop and its location within the lens. The entrance pupil is the image of the aperture stop as it is seen when looking from the object space into the lens. The exit pupil is the image of the aperture stop as it is seen when looking from the image space into the lens. In the case where the aperture stop is seen directly, without looking through any lens elements, then the corresponding pupil will coincide with the aperture stop. A typical relationship of the entrance and exit pupils to the aperture stop location is illustrated in Fig. 4.2. Note first that there are three rays traced through the lens to the image. These three rays are often referred to in the analysis of a lens system. They are: first, the *optical axis*, which is a ray that passes through the center of all optical elements and the image surface; second, the *marginal ray*, which is a ray parallel to the optical axis that passes through the outer edge (the margin) of the aperture stop; and third, the *chief ray*, which is a ray from any off-axis object point that passes through the center of the aperture stop. From the top of Fig. 4.2 it can be seen that extending the chief ray from object space to the point where it intersects the optical axis determines the axial location of the plane of the entrance pupil. Then, extending

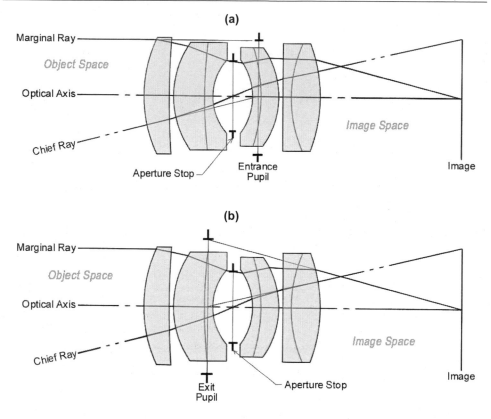

Figure 4.2 The image of the aperture stop as seen from object space is the entrance pupil (a). The image of the aperture stop as seen from image space is the exit pupil (b).

the marginal ray from object space to the point where it intersects the plane of the entrance pupil determines the diameter of the entrance pupil. Likewise, from the bottom of Fig. 4.2 it can be seen that extending the chief ray from image space to the point where it intersects the optical axis determines the axial location of the plane of the exit pupil. Then, extending the marginal ray from image space back to the point where it intersects the plane of the exit pupil determines the diameter of the exit pupil. Note that the term "extending" in this case indicates a straight line that is not refracted by any of the optics.

Similar to but totally unrelated to the aperture stop is the *field stop* of the lens or optical system. The field stop is the component that limits the field of view that will be seen or imaged by the optical system. We have stated that for the thin lens shown in Fig. 4.1, the detector has a diameter of 25 mm. This detector might be the input surface of an image intensifier or a charge-coupled device (CCD) array within a video camera. Whatever the case, it is this detector that limits the field of view of the system. Therefore, it is the system field stop.

Were the detector actually a CCD array within a TV camera, it would follow that the final image produced by the system (lens + camera = system) would not be 25 mm in diameter, but would most likely be a rectangle with a height-to-width ratio of 3:4 and a diagonal dimension of 25 mm. While not affecting the lens design process significantly, this rectangular image format would lead to the field of view of the system being specified as 8-deg horizontal by 6-deg vertical, rather than simply 10 deg. While the system detector is often the component that limits the field of view and, as a result, the system field stop, there are cases where the detector is capable of sensing a larger field of view than the optics deliver. In this case, the field stop will be a physical surface within the system. That field-stop surface must be precisely located at an internal image surface within the optical system.

4.4 Reference Coordinate System

In discussing and describing optical systems it is important that we become familiar with the basic reference coordinate system that is being used. Figure 4.3 illustrates the most common system, the one that is used throughout this book. It can be seen that there are three mutually perpendicular axes that are designated X, Y, and Z. The Z axis is also referred to as the optical axis, or centerline of the system. In most cases, the subject, all optical elements, and the image will be centered on and perpendicular to the Z axis, which is typically assumed to be horizontal. The Y axis is any vertical line that intersects the Z axis at a right angle. The Y-Z plane is a plane that contains both the Y and the Z axes. The Y-Z plane is referred to as the *tangential* plane. The X axis is any horizontal line that intersects the Z axis at a right angle. The X-Z plane is a plane that contains both the X and the Z axes. The X-Z plane is also referred to as the *sagittal* plane.

When it comes to the object of sign (+ or −), the most important consideration is consistency throughout your work. Generally, and throughout this book, distances along the Z axis that take us farther to the right, or toward the

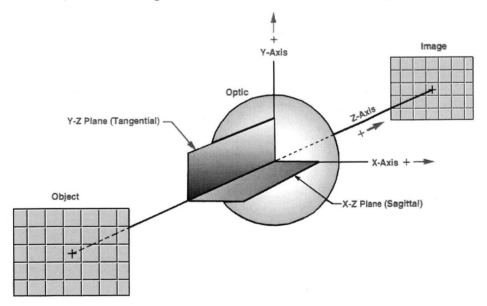

Figure 4.3 The standard reference coordinate system, as used in this and most contemporary optical texts.

image in the simplest case, are designated positive. In the case of the Y axis, distances away from the Z axis in an upward direction are designated positive. For the X axis, distances away from the Z axis to the right (as seen from the object) are designated positive. This, then, is the most basic and common coordinate system. The introduction of mirrors, prisms, tilted and/or decentered surfaces will obviously complicate things considerably. This presentation will be limited to the more simple and familiar configurations.

4.5 Thin-Lens Power

The *power* of a lens is an indicator of how much that lens will converge or diverge an incident wavefront of light. Lens power is typically indicated by the symbol Φ, and it may be specified in several ways. Most commonly, lens power is given as the reciprocal of the EFL. For example, the power of the lens shown in Fig. 4.1 would be $f = 1/142.9 = 0.007$. A second method of lens-power specification, used most often for camera close-up lenses and for eyeglasses, is to specify the lens power in diopters (D). This is equal to the reciprocal of the EFL in meters. For this example, where the EFL is $142.9/1000$ m, the lens power can be specified as $1/0.1429$ m $= 7.0$ D.

When several thin lenses are arranged such that their separations are negligible, then the powers of those lenses may be added directly. This is a very useful characteristic, considering the fact that it is not possible to determine the focal length of a combination of thin lenses by merely adding their focal lengths. This is illustrated in Fig. 4.4, where two examples are shown. When there are two positive 100-mm EFL thin lenses with a small separation, the EFL of the combination can be found by adding their powers and then taking the reciprocal of the sum. In this case the power of each thin lens will be $\Phi = 1/100 = 0.01$. The sum of the two would be $\Phi_{\text{combined}} = 0.01 + 0.01 = 0.02$, and the combined EFL will be $1/0.02 = 50$ mm.

The second case, shown in Fig. 4.4, deals with a common situation where the two thin lenses have powers of different sign. In this case we have a thin positive lens with an EFL of 50.0 mm, and a thin negative lens with an EFL of -100.0 mm. Adding the two lens powers we have $\Phi_1 + \Phi_2 = (0.02) + (-0.01) = 0.01$ for the combined power of the two lenses. The combined EFL would then be $1/0.01 = 100$ mm.

The calculations become just a little more complex when the space between the thin lenses is increased to some significant dimension. The following formula is used to calculate the combined power of two thin lenses with a significant separation:

$$\Phi_{\text{combined}} = \Phi_1 + \Phi_2 - (d\Phi_1\Phi_2),$$

where Φ_1 and Φ_2 are the individual lens powers and d is the lens separation.

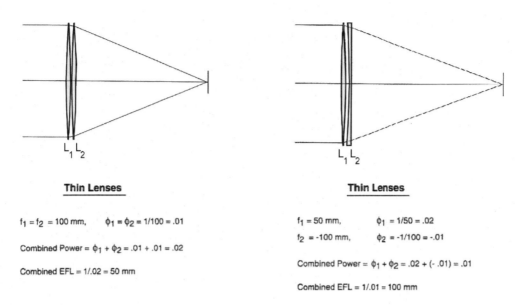

Thin Lenses

$f_1 = f_2 = 100$ mm, $\phi_1 = \phi_2 = 1/100 = .01$

Combined Power = $\phi_1 + \phi_2 = .01 + .01 = .02$

Combined EFL = $1/.02 = 50$ mm

Thin Lenses

$f_1 = 50$ mm, $\phi_1 = 1/50 = .02$

$f_2 = -100$ mm, $\phi_2 = -1/100 = -.01$

Combined Power = $\phi_1 + \phi_2 = .02 + (-.01) = .01$

Combined EFL = $1/.01 = 100$ mm

Figure 4.4 Thin-lens power Φ is a valuable concept, in that it permits the combined power and focal length of a combination of thin lenses to be easily determined.

In a case of this type it will be easier to use the formula that deals directly with the EFLs of the two lenses and the separation d between them. For example, looking back at the lens that was shown in Fig. 4.1, the system specifications that were given called for a 10-deg field of view and a lens speed of $f/2.0$. An experienced optical engineer would conclude from these numbers that a Petzval lens form will be required in order to ensure a final lens design with adequate image quality. A Petzval lens can be represented by two positive thin lenses separated by a substantial airspace. In the classic Petzval lens layout, the EFL of the first element is twice the EFL of the combination, while the EFL of the second element and the lens separation are both equal to the EFL of the combination. Since we have established in Fig. 4.1 that the required lens EFL is 143 mm, it follows from the preceding description that the EFL of the first thin lens will be 286 mm, while the separation between the lenses and the EFL of the second thin lens will both be equal to 143 mm. This arrangement for a thin-lens layout of an $f/2.0$, 143-mm EFL Petzval lens is shown in Fig. 4.5.

In addition to the power-based formula given earlier, Fig. 4.5 contains two formulas that use the individual thin-lens EFLs. The first of these permits the direct calculation of the EFL for the two-lens combination. The second formula permits calculation of the distance from the second thin lens to the image, i.e., the back focal length (BFL). Given the focal lengths of the two thin lenses (f_1 and f_2), and the distance d between lenses, the following calculations yield both the combined EFL and the BFL:

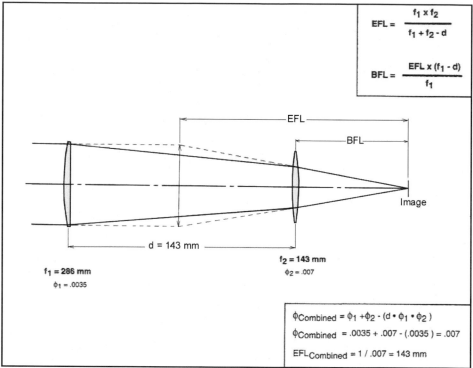

Figure 4.5 When two thin lenses are separated by a substantial distance *d*, the focal length of the combination (EFL) and the back focal length (BFL) can be calculated using the formulas shown here.

$$\text{EFL} = \left(f_1 \times f_2\right)\big/\left(f_1 + f_1 - d\right)$$
$$\text{EFL} = \left(286 \times 143\right)\big/\left(286 + 143 - 143\right) = 143 \text{ mm}$$
$$\text{BFL} = \text{EFL} \times \left(f_1 - d\right)\big/ f_1$$
$$\text{BFL} = 143 \times \left(286 - 143\right)\big/ 286 = 71.5 \text{ mm.}$$

4.6 Ray Trace of a Thin Lens (Object at Infinity)

The basic simplicity of the thin lens concept makes ray tracing through a thin lens equally simple. This ray tracing may be done either graphically or mathematically, with equal ease and effectiveness. First, as we have seen, the optical axis represents the path of a ray traveling from the center of the object to the center of the image. There are just two additional rays that need to be traced in order to establish all of the required thin lens characteristics: these two rays are commonly referred to as the *marginal ray* and the *chief ray*. The marginal ray is that ray that starts at the center of the object and passes through the outermost edge of the optical system's aperture stop. The chief ray is that ray that starts at the outer edge of the object and passes through the center of the optical system's

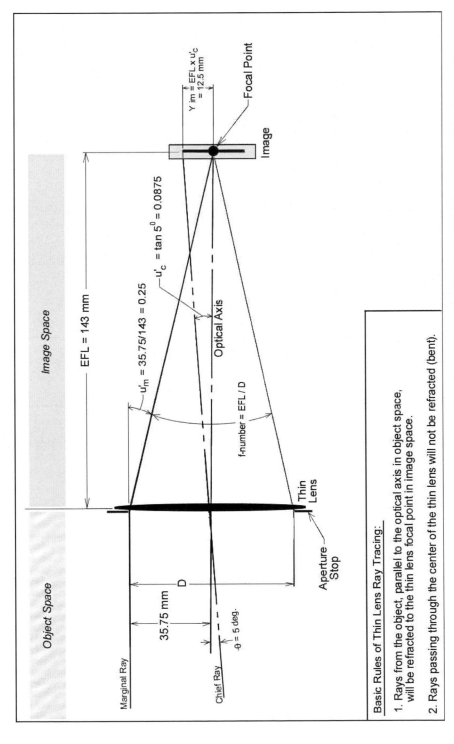

Basic Rules of Thin Lens Ray Tracing:

1. Rays from the object, parallel to the optical axis in object space, will be refracted to the thin lens focal point in image space.

2. Rays passing through the center of the thin lens will not be refracted (bent).

Figure 4.6 Ray tracing through a thin lens with the object at infinity. Given the EFL, the half-field angle, and the aperture diameter, the marginal and chief rays can be traced as shown to determine other thin-lens characteristics. Ray tracing may be done graphically or by the application of basic trigonometric formulas.

aperture stop. In the case illustrated in Fig. 4.6, the object is assumed to be at infinity. As a result, all rays originating at the center of the object, including the marginal ray, can be drawn parallel to the optical axis as they approach the thin lens. The chief ray is the ray that passes through the center of the aperture stop after approaching the thin lens at the maximum specified field angle of 5 deg. In setting up a single thin lens for graphic ray tracing it can be assumed that the aperture stop is in coincidence with the thin lens unless it has been otherwise specified. The EFL of this thin lens is specified to be 143 mm. The graphic ray-trace layout should include the focal point of the thin lens. In this case that would be a point on the optical axis 143 mm to the right of the thin lens. By definition, all light rays from the center of the object at infinity will approach the lens traveling parallel to the optical axis, and they will be refracted by the thin lens such that they all pass through this focal point.

The height if the marginal ray at the lens will be equal to one half the aperture-stop diameter. Again, it has been specified that the aperture-stop diameter in this case is 71.5 mm. The height of the marginal ray at the thin lens will be equal to the aperture stop radius, which is $71.5/2 = 35.75$ mm. The angle that the chief ray makes with the optical axis as it approaches the lens is important. Ray angles in object space are traditionally designated as u, while ray angles in image space are designated as u'. From Fig. 4.6 we see that the angle of the marginal ray in object space (u_m) is zero, while for the same ray in image space, its slope angle is:

$$u'_m = 35.75/143 = 0.25.$$

While it might seem reasonable to convert this angle to degrees (about 14 deg), that will not be necessary, since all subsequent ray tracing and calculations will use the value in its present form. When, as is most often the case, the image is formed in air, the slope angle (u'_m) of the marginal ray in image space is referred to as the numerical aperture (NA) of the lens. This NA is a term most often used when describing the characteristics of finite conjugate optics, such as a microscope objective. A familiar formula, and one worth remembering, relates the NA to the $f/\#$ of a lens:

$$f/\# = \frac{1}{(2\text{NA})}.$$

In this case:
$$f/\# = \frac{1}{(2 \times 0.25)} = 2.0.$$

Continuing with the thin-lens ray-trace discussion, the *image plane* is a plane that is perpendicular to the optical axis, and it is located at the focal point of the thin lens. The size of the image is determined by the intersection of the chief ray with the image plane. Because in Fig. 4.6 we have assumed the aperture stop to be in coincidence with the thin lens, the chief ray will pass through the center of the thin lens at the optical axis. Any ray passing through the center of a thin lens will not be refracted; that is, its angle with the optical axis will not change as it passes through the thin lens from object space to image space. In the case shown in Fig. 4.6, the chief ray is traveling from the object to the center of the thin lens at an angle of 5 deg with the optical axis; thus, the angle u_c for the chief ray equals the tangent of 5 deg = 0.0875. This angle will be constant as the chief ray travels from the object to the image. The height of the chief ray as it intersects the image plane may be found by multiplying its slope angle in image space u'_c by the EFL of the lens:

$$Y_{\text{image}} = 0.0875 \times 143 = 12.5 \text{ mm}.$$

The following two basic rules of thin-lens ray tracing are very useful, easy to work with, and easy to remember:

(1) A ray that is parallel to the optical axis in object space will be refracted by the thin lens to pass through the focal point.
(2) A ray that passes through the center of a thin lens will not be refracted. Its angle with the optical axis will be the same in both object and image space.

It is important to establish a sign convention for ray tracing and to maintain that convention consistently throughout the ray-trace analysis. The reference-coordinate system and sign convention shown in Fig. 4.3 are applicable to thin-lens ray tracing. All thin-lens ray tracing is done with rays that lie in the Y-Z (tangential) plane. The sign of the angles u and u' will be positive when the value of Y for the ray is increasing and negative when it is decreasing. Consider the angle of the chief ray as it appears in Fig. 4.6. At the far left, in object space, the Y value of the chief ray is negative (about −5 mm). At the lens this Y value has become zero. Since zero is greater than −5 mm, the Y value of the ray has been increasing, thus the sign of the 0.0875 chief ray slope angle is positive. For the marginal ray, the Y value at the lens is +35.75 mm. When the marginal ray reaches the image plane its value is zero. Since zero is less than +35.75, the Y value has been decreasing, thus the sign of the 0.25 marginal ray slope angle in image space is negative. A simple sketch of the thin-lens optical system will always be extremely helpful in visualizing all of these ray paths.

4.7 Ray Trace of a Thin Lens (Finite Object Distance)

When the object is at a finite distance from the thin lens, as shown in Figure 4.7, the thin-lens ray trace becomes just a little more complicated. Because the marginal ray (that ray originating at the center of the object and passing through the outer edge of the thin lens) is not parallel to the optical axis in object space, a new method must be employed to determine the axial location of the image plane. In Fig. 4.7, the same 143-mm, $f/2.0$ thin lens that was shown in Fig. 4.6 is illustrated, with the object now located at a finite distance of 500 mm. Following the same basic procedure established earlier, the first step is to trace the path of the marginal ray. That ray will, by definition, originate at the point where the object plane intersects the optical axis and then travel to the edge of the thin-lens aperture. In this case, the angle of that marginal ray can be calculated as $35.75/500 = 0.0715$. At this point, the marginal ray will be refracted by the thin lens. Earlier it was determined that an incident light ray, parallel to the optical axis, striking the thin lens at a height of 35.75 mm, will be refracted through an angle of –0.25. This will also be the case for our new marginal ray which is traveling at an angle of +0.0715 when it strikes the thin lens at that same height. The resulting marginal ray-slope angle following refraction will be $u' = 0.0715 - 0.25 = -0.1785$. The new image distance may now be calculated as:

$$\text{Image distance} = 35.75 \text{ mm}/0.1785 = 200 \text{ mm}.$$

When considering a second approach to finding the image distance, it is possible to trace two rays originating at the same off-axis object point. Assuming that object point to be 25 mm below the optical axis (see Fig. 4.7), we know that the chief ray will pass through the center of the thin lens without being refracted. We can see that the angle of this chief ray ($u = u'$) will be $25/500 = -0.05$ as it travels from object to image. If a second ray from that same object point is traced parallel to the optical axis in object space, we know that it will be refracted by the thin lens such that it passes through the focal point of the lens. This ray angle in image space will then be $u' = 25/143 = 0.175$.

We have now traced two rays from the same object point and determined that they are traveling through image space at angles of 0.050 and 0.175. The difference between these two angles ($0.175 - 0.050 = 0.125$) represents the rate at which these two rays are converging toward each other, i.e., the angle between the two rays. We also know that, at the thin lens, these two rays are separated by 25 mm. Based on the convergence angle of 0.125, we can now determine that the two rays will intersect at a distance of $25/0.125 = 200 \text{ mm}$ to the right of the thin lens. This intersection point establishes the location of the thin-lens image plane. This is equal to the value found previously by tracing the marginal ray.

Having determined the object and image locations, we can now calculate the actual sizes of the object and image. Earlier we had arbitrarily assigned an object height of –25 mm, and through ray tracing of the chief ray, we can see that this will result in an image height of $200 \times 0.05 = 10.0 \text{ mm}$. If the object is a 50-mm diameter illuminated disk, then the image will be a corresponding disk with a diameter of 20 mm. From these data we can compute the magnification of this lens in this configuration to be $20/50 = 0.40\times$.

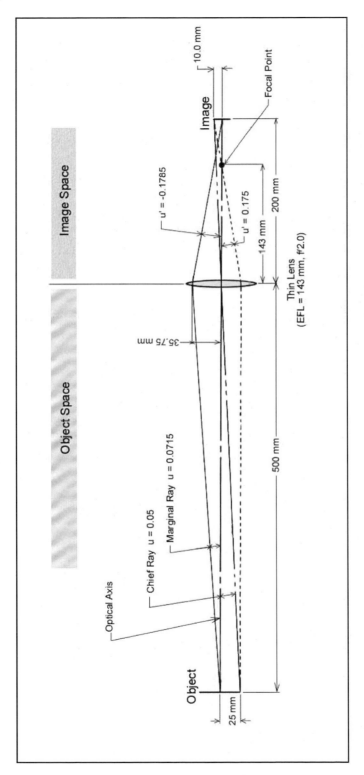

Figure 4.7 Shown is a ray trace through a thin lens with the object at a finite distance. The marginal ray is traced from the object to the image to determine the image distance. In a second approach, the chief ray and a second ray (dotted) from the same object point and parallel to the optical axis in object space are traced as shown to yield the same result.

If, on the other hand, we assume the earlier condition where the image plane contains a detector with a diameter of 25 mm, then we can conclude that a disk in the object plane that is $25/0.40 = 62.6$ mm in diameter will be imaged by the lens onto that detector. This 62.6-mm diameter object disk is referred to as the *field of view* of this thin-lens system.

The magnification factor can also be found by comparing the ratio of the image distance to the object distance (200/500 = 0.40×). A less direct, but often useful method of determining magnification involves comparison of the marginal ray angle in object and image space In this case, the minus sign correctly indicates an inversion of the image relative to the object.

To permit the calculation of image brightness, it is necessary to determine the *f/#* of the light bundle being focused at the image plane. In this case, that *f/#* can be found by dividing the image distance by the thin-lens diameter:

$$f/\# = \frac{200}{71.5} = f/2.8.$$

We might also use the formula for the *f/#* established earlier using the marginal ray angle in image space, i.e., the numerical aperture

$$f/\# = \frac{1}{(2\text{NA})} = \frac{1}{(2 \times 0.179)} = f/2.8$$

Either approach works equally well and should produce the same numerical result. Whenever two approaches are available, it is wise to employ both to check for the possibility of error. It is interesting to note that, for this example where we have a finite object distance, even though we are dealing with an *f*/2.0 lens, the *f/#* at the image plane is found to be *f*/2.8.

4.8 Rounding Off

It may have been noticed throughout this chapter that numerical results have frequently been rounded off in what may have appeared to be a random fashion. Legitimate rounding off has indeed been done in order to clarify and simplify the presentation, but it has not been done randomly, or without thought. It is always important to maintain a level of precision that is consistent with the requirements of the optical system being analyzed. In nearly all cases, the six-place calculations can and should be left to the computer. A rule of thumb that has served me well in most instances has been to round off such that the error introduced is less than 1%. For example, at the beginning of this chapter we were given a detector size of 25 mm and a required field of view of 10 deg. Using the maximum image height of 12.5 mm and the half-field angle of 5 deg, we find the required EFL of the thin lens to be 142.876 mm. This was rounded up to 143 mm to simplify subsequent calculations. The error introduced by doing so was less than 0.1%. A reality check at this point will tell us that if we were to purchase or manufacture a typical 143-mm lens, the standard tolerance on its EFL would be in the +/–1% range.

In the calculations shown in Fig. 4.5, the individual lens powers are found to be 0.0035 and 0.0070. Here, we might be tempted to round the first number to 0.0040, until a quick calculation reveals that this introduces a 14% error . . . clearly not acceptable. In Fig. 4.6, the initial value for the tangent of the 5-deg half-field angle is given as 0.0875. This might have been rounded to 0.088. While this would represent a round-off error of just 0.6%, it was deemed unacceptable because this is one of the very basic system specifications.

When it comes to rounding off, hard and fast rules are not possible, nor are they desirable. The decision to round off, or not, should always be based on a sound reason. If a sound reason is not apparent, do not round off. The type of conflicting results that might occur due to rounding off can be seen by comparing results from Figs. 4.7 and 4.8. The first calculation of image distance resulted in a value of 200 mm. Knowing the object distance to be 500 mm and the thin-lens EFL to be 143 mm, we can use the formula given in Fig. 4.8 to calculate the image distance:

$$S' = (f \times S)/(S - f) = 71500/357 = 200.28 \text{ mm.}$$

The difference between these two results is 0.14%, a negligible and perfectly acceptable round-off error.

We must keep in mind the fact that these are thin-lens calculations, which have been introduced as approximations in their own right. To attempt to accomplish an extreme degree of precision here would represent a complete waste of time and energy. Better that time and energy be devoted to assuring that the correct formulas are being used and to developing a comfortable sense of how the ultimate lens will perform and what its basic characteristics will be.

4.9 Thin-Lens Formulas

Yet another approach to thin-lens analysis involves the use of established, traditional, thin-lens formulas rather than graphical or mathematical ray tracing. Figure 4.8 illustrates a generic thin-lens configuration with the object at a finite distance from the lens. In a typical case, the EFL of the lens and the object distance might be given. After applying the Gaussian formulas in the left-hand box, it is possible to determine the corresponding image distance. The example we used earlier involved a 143-mm EFL thin lens f with an object distance of 500 mm. The formula for S' can be used to determine the image distance:

$$S' = (f \times S)/(S - f) = (143 \times 500)/(500 - 143) = 200.3 \text{ mm.}$$

On another occasion the total object-to-image distance might be known, along with the required magnification factor. In that case, the goal would be to

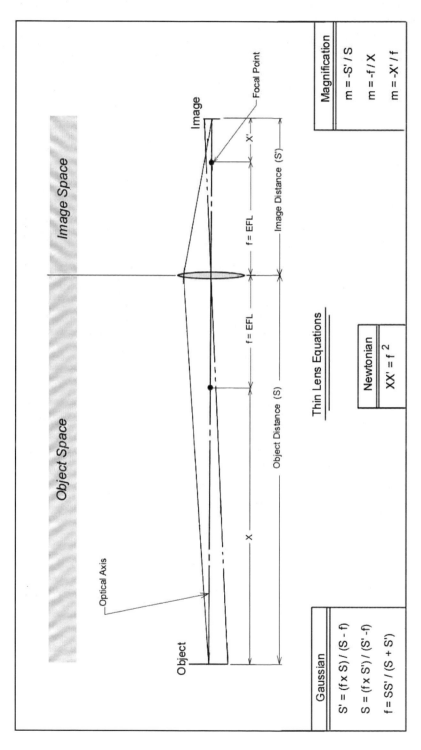

Figure 4.8 In lieu of graphical or mathematical ray tracing, the formulas shown here can be used to determine thin-lens parameters, including object distance, image distances, and magnification factors.

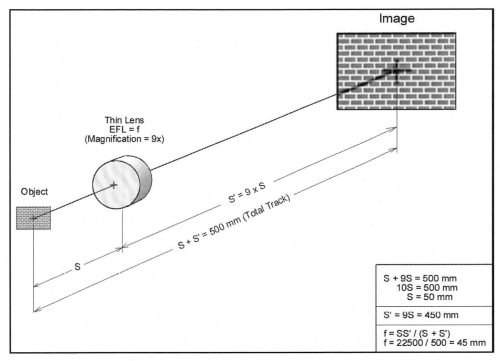

Figure 4.9 Diagram showing the application of thin-lens formulas to determine the required thin-lens EFL, given the total track and required magnification.

determine the required thin-lens focal length. Assume, for example, the case illustrated in Fig. 4.9. Here we have a system with a 35-mm transparency (a slide) as an object and a rear projection screen at the image plane. The distance from the slide to the screen (total track) is fixed at 500 mm, and we wish to produce an image on the screen that is magnified by a factor of 9×. Since S' must be equal to $9S$, it follows that $S + S' = 10S = 500$ mm. Therefore, S must be equal to 50 mm. Knowing that $S = 50$ mm and $S' = 9S = 450$ mm, we can find the focal length f of the required thin lens using the following formula:

$$f = (S \times S')/(S + S') = 22500/500 = 45 \text{ mm}.$$

Knowing the object size to be 24 × 36 mm, we can determine that the image on the screen will be 9× that, or 216 × 324 mm.

These formulas dealing with S and S' have become known as the Gaussian thin-lens formulas. The equation shown in the center box in Fig. 4.8 ($XX'=f^2$) is known as the Newtonian thin-lens formula. This formula is easier to remember and it is also somewhat simpler algebraically. It should be noted that in the Newtonian formula the terms X and X' are distances along the Z axis, from the lens focal point to the object and image. In this case, the use of the term X is based on tradition; it is in no way related to the X axis as it has been defined in our local coordinate system.

The Newtonian formula can be used to solve the example discussed earlier, where the lens has a focal length f of 143 mm and the object distance S is 500 mm. Using these numbers, we can compute the value for X:

$$X = S - f = 500 - 143 = 357 \text{ mm.}$$

Knowing f and X, we can now find X':

$$X' = f^2 / X = 143^2 / 357 = 57 \text{ mm.}$$

The new value for S' will then be:

$$S' = f + X' = 143 + 57 = 200 \text{ mm.}$$

This confirms our earlier calculation done using the Gaussian formulas and thin-lens ray tracing.

4.10 Applications of Thin-Lens Theory

A few examples will be helpful in demonstrating the value and usefulness of thin-lens ray tracing. Assuming once again our basic 143-mm EFL f/2 lens, the question might arise as to over what range of object distances this lens would provide satisfactory image quality. This first question leads immediately to a second. That is, what is the precise definition of satisfactory? Typically, we might be told that any conditions where the blur-spot diameter, due to focus error, is less than 0.02 mm would be considered satisfactory. For the purpose of this example, we will ignore lens aberrations and assume that the lens forms a perfect image at the thin-lens focus. For an object at infinity, this perfect image will be formed at the focal point of the thin lens (see Fig. 4.10). As the object moves from infinity to a point closer to the lens, the point of best focus will move to the right of the focal point. If our detector is fixed at the focal point, it can be seen that the change in focus will result in a finite blur spot at the detector. Because this is an f/2 light bundle, that blur spot will reach a diameter of 0.02 mm when the amount of defocus reaches 0.04 mm.

The object distance that corresponds to this 0.04-mm focus shift can be found using the Newtonian thin-lens formula:

$$XX' = f^2$$
$$X = f^2 / X' = 143^2 / 0.04 = 511225 \text{ mm } (0.51 \text{ km}).$$

From this result we conclude that, if the lens is focused at infinity, then all objects at distances from 0.51 km (about ⅓ of a mile) to infinity will be in satisfactory focus.

There is one additional step that we might take in a case like this. If we move the detector 0.04 mm to the right, light from objects at a distance of 0.51 km will

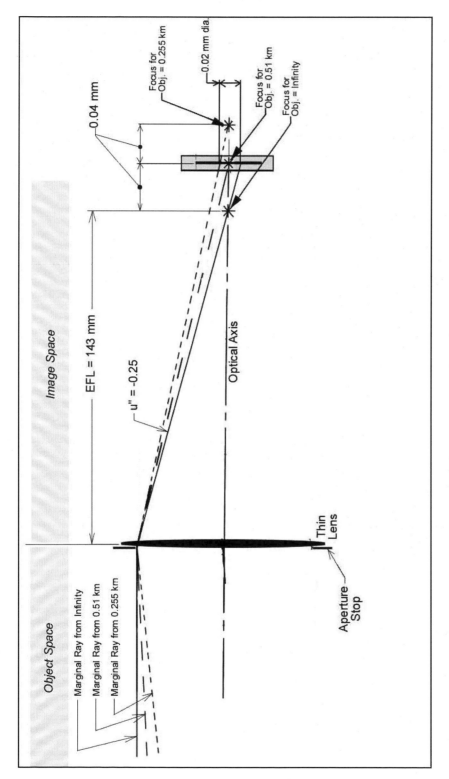

Figure 4.10 Thin-lens ray trace to determine the range of object distances over which satisfactory image quality will be produced. The midfocus object distance (0.51 km) is referred to as the hyperfocal distance.

be in sharp focus. In addition, we find that light from all object points from 0.51 km to infinity will be in satisfactory focus. Also, there will be a range of object distances inside 0.51 km that will produce a satisfactory focus condition. Calculations will reveal that range will be from 0.51 km down to 0.255 km. Thus, we can conclude that with the focus set to the 0.51-km object distance, satisfactory image quality will result over the range of object distances from infinity down to 0.255 km. The midvalue (0.51 km) is referred to as the *hyperfocal distance*. When lens focus is set to the hyperfocal distance, the lens will image all objects from half that distance to infinity with satisfactory image quality, i.e., the blur-spot diameter due to focus will be less than 0.02 mm.

This exercise is typical of the preliminary thin-lens calculations that are the responsibility of the optical engineer. The results of this type of work can be quite valuable. For instance, in this case they might allow the system designer to conclude that a focus mechanism for this lens will not be required. This could result in a substantial reduction in the overall cost and complexity of the final lens system.

Consider a second, somewhat more complex problem that can again be solved by applying the basic thin-lens formulas that are presented in this chapter. Figure 4.11(a) shows a two-element Petzval-type thin lens, similar to the one in Fig. 4.5. This example will deal with the question of how that lens might be designed so that it is capable of being continuously focused from infinity down to an object distance of 2000 mm. The most obvious approach is to make the space between the second lens and the detector (image) adjustable. For a first approximation, we can assume an object distance of 2000 mm as the X quantity in the thin-lens formula: $XX' = f^2$. Using the 2000-mm value for X and 143-mm value for the EFL f of the lens, we can calculate the value for X':

$$XX' = f^2$$
$$X' = 143^2/2000 = 10.2 \text{ mm.}$$

It is possible to reverse the lens layout, add the 10.2 mm to the nominal BFL, assume the detector to be the object, and then trace rays from that object to the final image [see Fig. 4.11(b)]. This ray trace will be done in two steps, using the thin-lens formula:

$$S' = (S \times f)/(S - f).$$

First, for the 143-mm thin lens:

$$S' = (81.7 \times 143)/(81.7 - 143) = -190.6 \text{ mm.}$$

The minus sign indicates that the image of the detector, as formed by the 143-mm thin lens, will be to the left of that lens. This image can now be used as the object to ray trace through the 286-mm thin lens to find the ultimate image

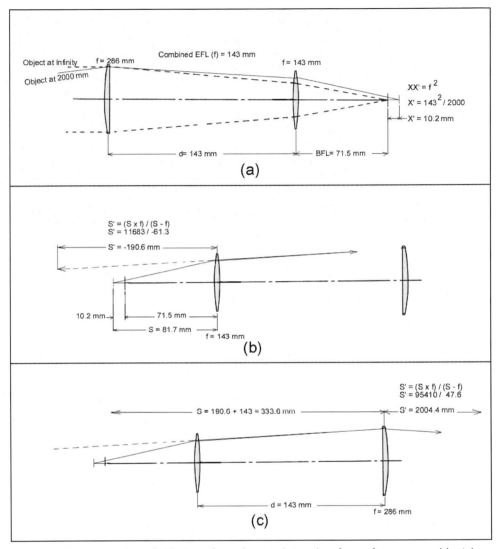

Figure 4.11 Application of thin-lens formulas to determine focus for a near object in a Petzval-type lens. (a) The approximate value of 10.2 mm is found using the thin lens formula. Parts (b) and (c) show the thin-lens ray trace, which confirms the adequacy of the initial approximation.

location for the Petzval lens combination. The object for the 286-mm lens is located 333.6 mm to its left [see Fig. 4.11(c)]. Applying the preceding formula once again, we have:

$$S' = (333.6 \times 286)/(333.6 - 286) = 2004.4 \text{ mm}.$$

This result is an acceptable approximation of the 2000-mm object distance that was our design goal (error <0.25%).

After completing a perfectly reasonable solution to the focus problem, it is often the engineer's fate that someone will come up with a different, and perhaps better, approach that will invalidate the initial solution. In this case, it may be felt that moving the entire lens assembly relative to the detector is a less-than-optimum approach. Rather, one might ask if it is possible to achieve the desired focus capability by moving just the rear lens element while maintaining a constant relationship between the front lens and the detector. This sounds like a pretty good approach . . . let's see if it can be done. While a more elegant and direct approach is possible using concepts and tools that have not yet been presented, we will execute this example in a basic manner using only those thin-lens concepts and formulas introduced earlier in this chapter.

Again, in Fig. 4.12 we will be ray tracing the Petzval lens in reverse fashion, i.e., from the detector (image) to the object surface. We will start by moving the rear lens by 10 mm (a guesstimate based on earlier work), away from the detector and closer to the front lens [see Fig. 4.12(a)]. Using the thin-lens formulas shown, we can calculate the distance S' from the rear thin lens to the image of the detector formed by that lens. As shown in Fig. 4.12(a), that image distance will be 189.5 mm to the left of the rear lens. We can then calculate the image distance S' for the image of the detector as formed by the front lens. That distance is found to be 2527 mm to the right of the front lens. Since our objective here is to focus down to 2000 mm from the front lens, it is clear that the 10-mm shift of the rear lens is not enough to achieve the near focus that is required. Because the ratio of the near focus distance just calculated to the required near focus is $2567/2000 = 1.26\times$, it would be reasonable to increase the initial 10-mm motion to $10 \times 1.26 = 12.6$ mm. In Fig. 4.12(b) it can be seen that moving the rear lens 12.6 mm to the right will result in the first detector image forming at a distance of 204.2 mm to the left of the rear lens. Additional thin-lens calculations [shown in 4.12(b)] then show that the final detector image will be located at a distance S' of 1969 mm to the right of the front element. While not precisely equal to the 2000-mm near-focus goal, this result is acceptable, and the 12.6-mm rear-lens focus travel would be recommended.

It might be noted, perhaps with some concern, that this result of 1969 mm does exceed our previously recommended 1% rule of thumb. It is always important to understand clearly the true meaning of the results generated. In this case, we stated at the outset that the goal was to have a lens assembly that was capable of focusing on objects from infinity down to 2000 mm. The design we settled on will focus down to 1969 mm if we move the rear lens 12.6 mm closer to the front lens. This is obviously not a problem. It merely provides a bit of extra travel that may well be required when all manufacturing and assembly tolerances are taken into account. All would agree that if the final object distance were greater than 2000 mm, by any amount however small, that would not have been acceptable.

In an earlier example (Fig. 4.10) we concluded that it might be possible to get by without a focus mechanism. In this case where we must focus down to 2000 mm, that will clearly not be an option. One question to be considered here

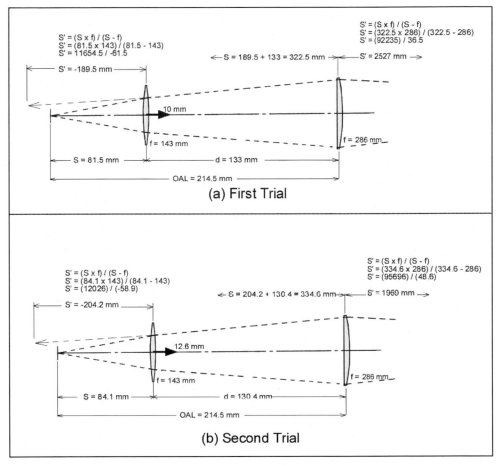

Figure 4.12 It is possible to adjust the focus of the lens assembly by moving only the rear lens element, while maintaining a fixed relationship between the front element and the image plane. Shown here are the two steps discussed in the text to determine the amount of rear lens travel required to focus from infinity down to an object distance of 2000 mm.

is whether the design should permit moving the entire lens relative to the detector (as in Fig. 4.11), or whether it will be better to maintain the lens detector relationship and achieve focus by moving the rear lens element only. While a bit more complex, the second approach does offer some advantages.

For example, in many lens designs the internal volume of the lens assembly will be purged and filled with dry nitrogen under slight pressure to prevent the accumulation of dirt and/or moisture on internal lens surfaces. If our design called for the entire lens to be moved away from the detector to accomplish focus, there would be a resulting increase in the internal volume of the lens. As this volume increase takes place, there would be a tendency for outside air to be drawn into the lens, along with dirt and moisture that might be in that air.

On the other hand, by using the second approach and moving just the rear lens element, we have the major advantage that the internal volume remains constant throughout the entire range of focus travel (see Fig. 4.13). Note from that figure that it is critical that we provide a port in the rear lens cell that will allow the dry nitrogen to flow between the front and rear volumes of the lens assembly as focus is taking place. It is through the understanding of basic design principles such as this that the optical engineer is able to better appreciate the importance of his work and, as a result, able to do a better job at it.

Whenever a design judgment such as this is implemented, it is important to consider all of the major tradeoffs that are involved. In everyday terminology, you hardly ever get something for nothing. In this case, we have gained the advantage of a design where the internal volume is constant during focus. In order to accomplish that, the separation between the two thin-lens elements has been varied by a substantial amount. It will be recalled from the thin-lens formula given in Fig. 4.5 that a change in lens separation will produce a change in EFL. It should be considered whether the amount of change involved here will have a negative impact on overall system performance. We can calculate the lens EFL when the focus is set to the near object distance of 2000 mm. For that case the lens separation d will be reduced from 143 mm down to 130 mm. The resulting thin-lens EFL will in turn be reduced from 143 mm down to 137 mm, a decrease of about 4%. While this is not a negligible amount, it would probably be deemed a reasonable tradeoff, considering the major design advantage that has been realized.

4.11 Mock-up of the Thin-Lens System

There is an intermediate step between thin-lens theory and final thick-lens design that often permits valuable information to be generated at very little cost in terms of time and money. I refer to this as the *simple lens mock-up stage*. We have stated that the thin lens is not a reality, it is a concept. In terms of actual hardware, the simple lens is the nearest thing to a thin lens. The simple lens is a single lens element made from standard optical glass. It will have a diameter and

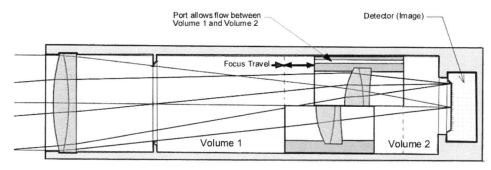

Figure 4.13 In order to prevent internal contamination, a lens assembly will often be sealed, purged, and filled with dry nitrogen under a slight internal pressure. Moving the rear lens element for focus maintains a constant internal volume.

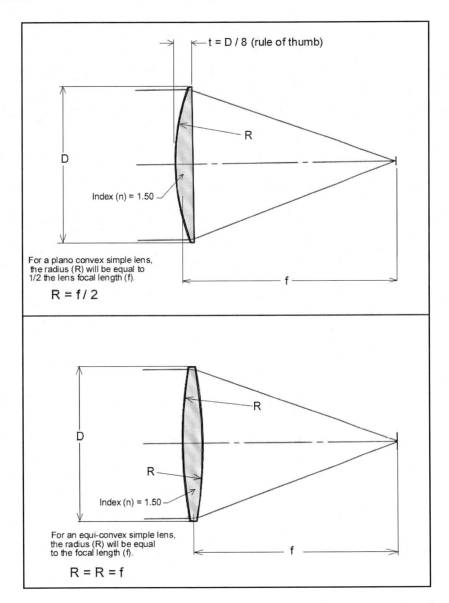

Figure 4.14 The thin lens can be simulated by a simple lens for purposes of preliminary computer analysis. The above approximations show the relationship between lens focal length *f* and the radii of the two most common forms of simple crown glass lens elements.

optical power quite close to that of the thin lens, while not dealing with the subject of image quality. Figure 4.14 shows the two most common and useful forms of simple lenses, the plano-convex and the equi-convex lenses. Basic relationships of lens radii and lens power are shown. There are two ways in which the optical system can be carried from the thin lens to the simple lens mock-up stage. First, it can be set up for computer analysis using one of several optical design software packages available today. Alternately, it is possible to purchase simple lenses and to physically mock-up the system on a simple lens bench. In this day and age, the first approach has become, by far, the preferred way to handle the problem.

The computerized simple lens mock-up requires that the designer have access to a basic personal computer system, including hardware and optical design software. While the experienced lens designer would most likely have access to such a system, it is becoming increasingly common and beneficial to provide a similar set of tools to all optical engineers and optical system designers. It is important that these personnel have the opportunity to become familiar with the basic procedures involved in the use of such computer software. While the optical engineer and system designer are not expected to generate final lens designs, the availability of modern computer systems is expanding their responsibilities to include preliminary lens design and analysis. While there are a number of suitable optical design software packages that are suitable for this work, examples presented here will assume the use of the OSLO-EDU program, which is available as a free download from the web site of its owner, Lambda Research.

For this exercise we will consider once again the Petzval lens form that is shown in Fig. 4.5. In order to simulate this lens using a computer system, we will first need to assign some basic physical characteristics to the two thin lenses we will be simulating. One bit of advice that I was offered to many years ago is worth repeating here: In the process of optical system design and analysis, a picture of the lens, as it is seen by the computer, will be worth much more than a thousand words (or numbers). The primary reason for this is that, in doing mathematical ray tracing, the computer does not consider the actual physical characteristics of the optical components. For example, a lens may have a negative edge or center thickness and the computer program will continue to happily trace rays through that lens. By drawing a picture of the optics, or better yet having the computer generate a drawing, it becomes readily apparent when an unacceptable physical condition has been assumed or generated.

For this example we will produce a computer model containing two simple lens elements similar to those that might be used to mock-up the thin-lens version of the Petzval lens. Because this is to be a 143-mm EFL lens with a speed of $f/2$, we can conclude that the first simple lens must have a focal length of 286 mm and a diameter of about 75 mm. There will then be a lens separation of about 135 mm (reduced from 143 mm to account for lens element thickness), followed by the second lens with a focal length of 143 mm and a diameter of about 60 mm. Assuming both lenses to be plano-convex, we can estimate the radius and thickness of the first lens based on the simple lens approximations that were given in Fig. 4.14. From this we can conclude that the radius on the first surface of the first lens will be equal to half its focal length, or 143 mm, and its thickness will be about one-eighth its diameter, or 10 mm. The material for this lens will be input as BK7, a common glass type with an index of about 1.52. Using the same approach, we can conclude that the second lens will have a radius of 72 mm, a thickness of 10.0 mm, and will also be made from BK7 glass.

All optical design software includes the ability to assign a value to the angle at which the marginal ray leaves a surface in lieu of a radius on that surface. The program will then automatically adjust the radius to produce that ray angle. This feature is known as an *angle solve*. The height of the marginal ray as it enters the first lens can be calculated to be one-fourth of the EFL, which is 35.75 mm. Because the EFL of the first element is 286 mm, the marginal ray must leave that lens at an angle of $35.75/286 = -0.125$. Because this is to be an *f*/2.0 lens, we can now calculate the angle of the marginal ray leaving the second lens to be $35.75/143 = 0.25$. We can assign these two angle-solve values to the rear surfaces of the two simple lenses generating the result shown in Fig. 4.15.

The distance from the second lens to the focal plane (the BFL) can also be determined automatically by the computer using what is known as a *height solve*. In this case, instead of a final thickness value at the last lens surface, we can specify the desired height of the marginal ray when it strikes the next surface. In this case we want that height to be zero. The computer software will then determine the thickness required to meet that constraint. The resulting simple lens system as input to the OSLO-EDU program is shown in Fig. 4.15. In addition to the two simple lenses, the marginal and chief rays are shown.

The diameter of the second lens has been set at 56 mm with a clear aperture of 54 mm. By working with this simple lens computer model, it is a simple matter to determine if this reduced diameter of the second lens will cause a serious reduction in the size of the off-axis light bundle (vignetting). OSLO-EDU

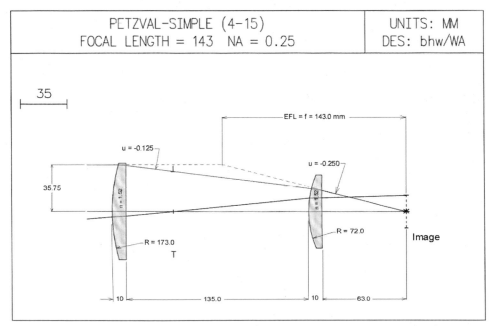

Figure 4.15 The 143-mm, *f*/2 Petzval lens developed earlier using thin-lens theory has been modified here to simple lens form for computer analysis. The simple lens approximations from Fig. 4.14 have been used to establish radii and thickness. Angle solves *u* have been used to establish the radii on second lens surfaces.

allows us to generate spot-diagram data for the maximum field angle. While actual image quality cannot be determined using simple lenses, we can determine quite accurately the amount of vignetting due to the reduced aperture of the second lens. A quick OSLO calculation tells us that there will be less than 10% vignetting at the maximum field angle of 5 deg.

With this model of the Petzval lens now in the computer, it is possible to revisit the focus problem that was covered earlier using thin-lens formulas. First, the angle solves must be removed so that established radii are not altered when we change the object distance. The height solve will be left on the last surface so that any change in focus due to change of object distance will be calculated automatically. When the object is at infinity, the BFL is found (by the height solve) to be 65.4 mm. When the object distance is reduced to 2000 mm, the resulting BFL is found to be 75.6 mm. This 10.2-mm focus shift coincides exactly with the result found by both the Newtonian thin-lens equation (Fig. 4.8) and the thin-lens ray-trace calculations (Fig. 4.11).

Next, we can simulate the example where lens focus is accomplished by moving only the rear lens element. In the thin-lens calculation we settled on a lens travel of 12.6 mm, which made the object distance equal to 1969 mm. In the case of the simple lens mock-up, we find that the object distance of 1969 mm will be accomplished by moving the second lens 15.0 mm closer to the first lens. The program shows us immediately that this change will reduce the EFL of the lens from 143.0 to 137.3 mm. This 4% reduction in EFL is consistent with the results that were found earlier using thin-lens formulas.

While this exercise has generated a number of answers that were easily achieved using thin-lens formulas, it has demonstrated that the use of a simple computer model makes it possible to generate easily additional data that takes us one step closer to the final, thick-lens design. We were, for example, able to determine the amount of vignetting that would be present due to our choices of lens-element diameters. There are any number of other system parameter changes that can be easily entered into the computer model, i.e., wavelength, $f/\#$, and field of view, and the results can be analyzed instantly. The computer and associated software should be regarded as an extension of the optical engineer's arsenal of tools, not just for detailed lens design and analysis, but also for many aspects of preliminary lens design and system calculations.

4.12 Review and Summary

The thin-lens concept represents a method by which the lens designer or optical engineer can represent the final optical system without having to deal with the complications of a thick-lens configuration. Using the thin-lens approach, it is possible to establish all of the basic lens system parameters by applying simple ray-trace methods and formulas. Early thin-lens analysis can be used to establish the relationships of the aperture stop, the entrance pupil, and the exit pupil to the lens system. The question of image quality must be reserved for later, thick-lens design and optimization.

Each thin lens has a degree of optical power, which can be found by calculating the reciprocal of its focal length. Formulas and methods have been presented demonstrating how the power of several thin lenses can be combined. It has also been shown how the tracing of just a few rays through a thin-lens system will reveal significant information, especially in terms of lens diameter, power, angular field of view, image size, and image location. In lieu of ray tracing, it is possible to conduct similar thin-lens analysis using the Gaussian or Newtonian thin-lens formulas to establish object to image relationships. A number of typical examples have been presented, illustrating the application of thin-lens theory to the solution of real problems.

Finally, we have devoted the final section of this chapter to a discussion of how the modern PC system can be used by the optical engineer to execute all types of thin-lens system analysis. This is a very important concept that represents a revolutionary method for the optical engineer to apply readily available computing tools to the early stages of optical system design and analysis. Once the engineer has created a simple lens mock-up of the optical system on his computer, it is much the same as having an actual lens system set up on a lens bench in the optics laboratory, where any number of variables may now be conveniently introduced to the system, and resulting changes to related parameters can be observed and documented.

In the next chapter the reader will be introduced more thoroughly to the OSLO-EDU optical design software package. This software is particularly effective when used by the optical engineer for preliminary system design and evaluation. This software eliminates the need for the optical engineer to learn the details of a more complex, full-featured lens design program. In addition, the OSLO-EDU program facilitates the inclusion of a wide variety of commercially available, catalog optical components, making possible an even more effective simulation of actual optical elements within the computer environment.

Chapter 5
Optical Design Basics

5.1 Introduction

This chapter will introduce the reader to another aspect of today's computer-based methods of optical engineering and lens design. Until the early 1990s, the topic of lens design has been treated as a very specialized, almost mysterious subject, and it would have been confined to an isolated chapter in any book dealing with optical engineering. This chapter will introduce the concept of the *optical designer*, that is, the individual responsible for executing a combination of tasks that, in the past, have been accomplished individually by either the optical engineer or the lens designer.

In modern lens design, considerable overlap now exists between the areas of expertise of the optical engineer and the lens designer. The person responsible for doing the combined work of both the optical engineer and the lens designer might reasonably be referred to as an optical designer. This chapter will describe the work of the optical designer and demonstrate how recent technological developments, particularly in the areas of computer hardware and optical design software, make that new position possible.

5.2 Historical Perspective

This entire discussion will become more meaningful if a real timeframe is considered. This second edition was written in the year 2008 . . . I have been working in the field of optical engineering for more than 40 years. When "recent developments" are referred to, that would be those occurring in the years between 1990 and 2005.

Owing to the very unique nature of the work involved, the lens designer has, in the past, worked alone while generating and analyzing a new lens design. Once completed, the lens designer's work would then be passed along to the optical or systems engineer for further analysis and incorporation into the final system design. If, at any point, a lens-design-related question were to come up, that question would be referred back to the lens designer for resolution. In other words, the lens designer had little to do with the overall system design, while the optical (or systems) engineer had nothing to do with the details of the lens design. This all worked reasonably well, but it did result in a high degree of specialization, especially in the case of the lens designer.

More recently, revolutionary technological developments have taken place in several areas that have led to a breakdown in this specialization and a blending of many functions that are now shared by the optical engineer and the lens designer. While this concept is difficult to quantify precisely, the illustration shown in Fig. 5.1 shows graphically the general nature of these changes. With the exception of a few isolated areas that remain the domain of one or the other, the lens designer is now called upon to consider a number of systems related to aspects of the design, while the optical engineer is expected to share in some of the day to day lens design tasks, particularly in the areas of image quality, tolerances, and environmental analysis.

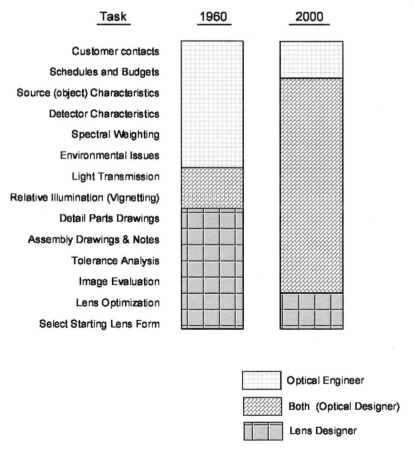

Figure 5.1 Until recently, the typical tasks within an optical design project have been clearly divided between the optical engineer and the lens designer. Today, with the advent of the personal computer and modern optical design software, the overlap between the two disciplines has increased dramatically.

The two principal technological advances that have made this blending of responsibilities possible are the development and growth of the personal computer and the availability of affordable optical design software that can be run on the PC. A historic review, from the perspective of this author, will be helpful in demonstrating the magnitude and significance of these developments.

My experience as a lens designer began with a major aerospace corporation in 1963. This operation involved the lease of an IBM-1620 computer that contained an integrated optical design software package. The hardware consisted of a computer with a typewriter-style printer and a separate card reader punch for the input and output of data. This was a fairly compact installation for the time, requiring a dedicated, air-conditioned computer room that was about 15×25 feet in size. Lens optimization was accomplished by control of third-order aberrations. Analysis of image quality was done by calculating aberration data and spot diagrams. Any plotting of aberration curves or lens drawings was done manually by the lens designer at his desk or drafting board. In terms of cost, this system was leased for about $2500 per month (equal to about $14000 per month in year 2000 U.S. currency). As another financial point of reference, that was about three times my monthly salary at that time. Later in the decade of the 1960s, the large mainframe computer became the machine of choice for lens design. Of course, the cost of such a machine was astronomical, making access a real problem. One could travel to a large computer center and rent time on a large computer, or one could access the computer via telephone lines through a time-sharing network. Looking back on these procedures, the inefficiencies involved were incredible as compared with today's methods. In the 1960s and 1970s, the optical-design software packages were significantly expanded and improved upon. More sophisticated lens-optimization techniques, including exact ray-trace data combined with high-order aberration coefficients, were now being incorporated. This made possible the design of lens systems with aspheric surfaces, as well as tilted and decentered surfaces. The advent of the plotter to produce hard copy of computer-generated graphics, including lens drawings, aberration curves, and spot diagrams, represented a significant work-saving addition to computer systems of that era.

In my experience, it was in the early 1970s that the mini-mainframe computer became available to the lens design department for their exclusive use for the design and analysis of optical systems. While considerably less expensive than a full-sized mainframe, the cost of these computers was still such that most of them were leased rather than purchased outright. At this point, the designer's input to the computer was still in the form of the 80-column IBM punched card. The efficiency of the mini-computer was such that several designers were able to share the system. As we entered the 1980s, computers were becoming much more affordable. Increased internal storage capacity and improved methods finally made it possible to eliminate the stacks of punched cards in favor of internal memory storage and CRT-keyboard interfacing. Later in the 1980s the personal computer (PC) became a viable platform for working lens design problems. Oddly enough, as the PC computing power and storage capacity increased, PC prices dropped rapidly into the very affordable range. Recognizing the great potential that existed, several providers of optical design software modified their programs, making them suitable to be run on the PC. Today, working as an independent optical engineering consultant, I find myself working on a PC system (including optical engineering software), giving me design and analysis capabilities that are infinitely greater than those found on the IBM-1620 computer that I used some 40+ years ago. Interestingly, the cost (in today's dollars) to purchase this system outright is just about equal to the $2500 monthly lease payment in the 1960s.

This is meant to be more than your typical "walked 6 miles to school" old-timer's story about how tough things were back in "the day." For example, today's automobile, like the computer, is considerably more sophisticated than the auto of the 1960s, yet its cost has increased by more than ten times during that same time period. In the case of the computer in general and optical design systems specifically, we have reached a point where performance has increased by many orders of magnitude, while the cost has come down to a point of absolute affordability.

5.3 OSLO Optical Design Software Package

On the subject of computer software for optical design, I would like to review the history of one specific product. This is obviously not the only product of this type, it is not the most powerful, nor is it the least expensive (except for the free version . . . more on that later). It is, however, a very good product, one that is representative of the present state of the art, and one with a history that demonstrates many of the significant developments in the field over the years. It is the software package that I have been using since 1992, and it will be used throughout the remainder of this book to generate design examples and to demonstrate design methods.

Optics Software for Layout and Optimization (OSLO) is an optical design software package that has been evolving for more than 30 years. Available from Lambda Research Corp., OSLO is one of the best-known and widely used optical design programs in the world today. Doug Sinclair, President and founder of Sinclair Optics, received his doctorate from the University of Rochester in 1965. After several years as a professor at the Institute of Optics, he established Sinclair Optics as a part-time venture in 1976. The company charter was to provide state-of-the-art optical design software to a rapidly growing group of users. Initially a very basic program that ran on early versions of the desktop computer, OSLO has developed over the years so that it now vies for the position of number one in the industry. In 2001, Sinclair Optics transferred responsibility for OSLO sales, service, and development to Lambda Research Corp., in Littleton, Massachusetts. Since that time, the staff at Lambda Research has continued to maintain OSLO as one of the leading optical design software packages available.

Recognizing trends in the 1980s, OSLO was one of the first professional-level optical design programs capable of being run using the Microsoft Windows operating system on the modern PC. This combination (OSLO, the PC, and

Windows) results in a program with total capability, combined with a previously unheard of level of user friendliness. This has precipitated the situation described earlier, where much of the mystery that had been associated with the lens design process has been swept away, making it possible for the optical systems engineer and many others working in related areas to perform a wide variety of basic optical design tasks. At the same time, this program has freed the lens designer of many tedious, time-consuming tasks, allowing him to broaden greatly the scope of the work in which he is involved. Certain aspects of tolerance analysis, environmental analysis, and many other areas of general optical engineering and system design have now been shifted to the position of shared responsibility between the optical engineer and the lens designer.

5.4 Introduction to Computer Design Using OSLO

OSLO-EDU is a free version of OSLO, intended primarily for educational purposes. Compared with the three other versions of OSLO that are also available, OSLO-EDU has a reduced number of features. It also restricts the user to working with optical systems that have up to but not more than 10 surfaces, including the object and image surfaces. OSLO-EDU has the basic ability to layout, edit, optimize, analyze, tolerance, and save a wide range of optical systems. While we will use OSLO-EDU here (whenever possible) to familiarize the reader with the basic tenants of computerized lens design and analysis, it must be understood that the limitations of this version of OSLO make it essentially mandatory that one of the advanced versions of OSLO, or any one of the other fine software packages available, be used for any serious design projects. Appendix B of this book contains reference and contact data for all of today's leading providers of optical design and engineering software.

I would like to suggest the following "plan" for the reader who would like to become familiar with the basic procedures used during computerized optical system design and analysis. First, you should read on to the end of this chapter, even though you may not be following along on your own computer . . . yet. Next, visit the Lambda Research web site (www.lambdares.com) and download the free version of OSLO-EDU. Next, download the *OSLO Optics Reference Manual*. After installing OSLO-EDU on your computer, get familiar with its on-line help system. This will be your basic OSLO-EDU operating manual. After these steps have been completed, along with some basic experimentation with your new system, you should be able to come back to this point and follow several of the more basic examples that are presented and discussed here.

The graphical interface provided by OSLO and Windows is the key to learning and applying the program efficiently. For example, when the program is initially opened for use, the window shown in Fig. 5.2 will be displayed. By clicking on one of the displayed buttons it is possible either to start a completely new lens, open the current lens (the lens that was being worked on when the program was last closed), open an existing lens (a lens selected from one of

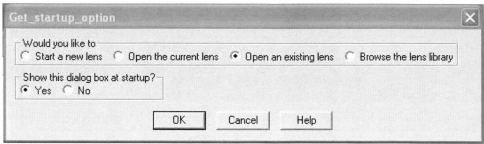

Figure 5.2 When OSLO-EDU is opened to run, the window above will be displayed, allowing the user to select a startup point.

several available lens libraries), or browse the OSLO-EDU lens library. If the "Open an existing lens" button is selected, the window shown in Fig. 5.3 will be displayed.

Assuming we will be doing some additional work on the Petzval lens that was discussed earlier, we will now double-click on the "petzval" icon at the bottom of the third column. This action will result in the desktop shown in Fig. 5.4. From this point the designer is able to do any number of things to modify and analyze this lens. For example, it will be noted from the lens title that this is a 50-mm focal length lens with a speed of $f/1.8$ and a half field of 7.5 deg. When the command to scale the focal length of the lens to 143 mm *(scl efl 143)* is entered, all lens curvatures and thicknesses will be scaled proportionately.

Open Lens File

Look in: edu

anamorph.len	hubble.len	pentprsm.len
anaprism.len	irfish.len	perfmag2.len
demotrip.len	lasrcomm.len	petzval.len
diodcoll.len	lasrdblt.len	prismirr.len
ebert.len	magnifyr.len	schmidt.len
grinrod-woodlens.len	mquartet.len	schwarz.len

File name: Open

Files of type: Lens files (*.len;*.osl)   Cancel

                                                                          Help

Library Directories: Private   Public

Figure 5.3 When the "Open an existing lens" startup option is selected, the window shown above will be displayed, allowing the user to select a lens for design and analysis.

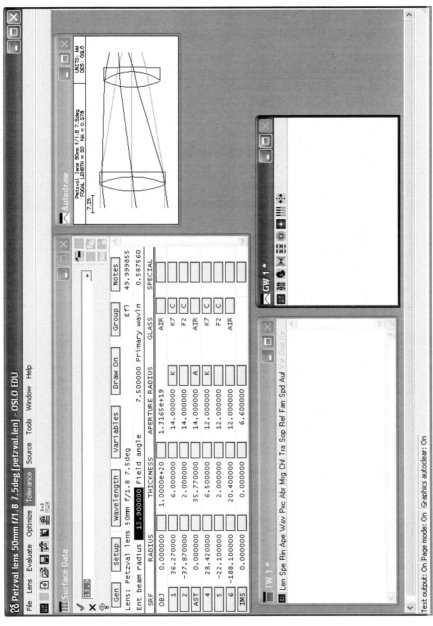

Figure 5.4 When the lens file named "petzval" is selected from the OSLO-EDU lens library, the desktop shown here will appear. This contains a lens surface data window, a lens drawing, a text window (TW1) and a graphics window (GW1).

The entrance pupil radius can then be changed to 35.75 and the (half) field angle to 5 deg. We next change the lens title to "Petzval Lens 143-mm $f/2$ 5 deg" and then save to lens file, with the name "petzval 143." The basic parameters of this lens are now the same as those of the simple lens model that was shown earlier in Fig. 4.15.

We can now add a second graphics window to our display and then display a lens picture in Window 1 and aberration curves in Window 2. We can also display a variety of lens data in the text window (see Fig. 5.5). The aberration curves show clearly that the image quality of this design is limited by residual field curvature (see ASTIGMATISM curves). We can see that residual on-axis aberrations are well balanced and that distortion and lateral color are also well corrected.

Previously we had estimated the amount of vignetting due to the reduced diameter of the second lens and concluded that it would probably be acceptable. Now, with a real lens model, we can confirm the actual amount of vignetting that will be present. From the surface data spreadsheet we can see that the last surface (SRF 6) has an aperture radius of 23 mm assigned. By tracing a bundle of light rays for a spot diagram at the maximum field angle, the analysis tells us that the amount of vignetting will be 7%. This is consistent with the 10% we had estimated earlier, and is clearly an acceptable amount.

Also, using the simple lens approach earlier, we had concluded that by moving the second lens about 15 mm closer to the first lens, while maintaining the distance from the first lens to the detector (image), we would accomplish the requirement of focusing down to a 2000-mm object distance. When we duplicate that exercise with this real lens example we find that, for a 15-mm shift, the corresponding object distance has been reduced to 1840 mm. This confirms the simple lens analysis and it represents a valid solution.

This analysis also shows that moving the rear lens 15 mm closer to the front lens will reduce the EFL of the lens assembly from 143 mm down to 136 mm. This represents a 5% reduction in the EFL, an amount that should not be problematic, but should be considered in subsequent lens system calculations.

Finally, Fig. 5.6 shows a solid model graphic of the Petzval lens and a set of modified modulation transfer function (MTF) curves from OSLO that allow a valid estimate of resolution for the on-axis case and for extended fields of view.

5.5 Laser Transmitting System

The following example will demonstrate methods that can be employed in using OSLO-EDU (or a similar software package) as a tool to design a basic optical system. This system will contain a projection lens and a collecting lens. The function will be to collect laser energy ($\lambda = 0.633$ µ) emitted from a 5-µm diameter optical fiber and then to focus that energy onto the input surface of a similar fiber at a distance of about 100 m. Figure 5.7(a) shows this system in simple lens form. The system will consist of two 60-mm EFL $f/2.0$ lenses. A ray trace of this system with two plano-convex lenses shows the spot diameter at the focus of the second lens to be about 50 µm, 10 times the required spot size. Being

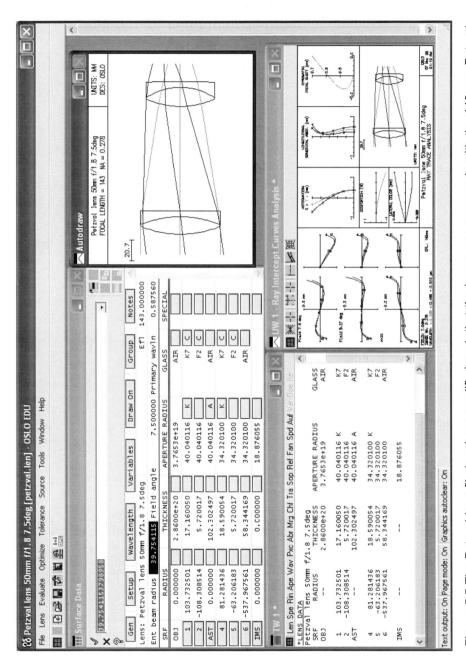

Figure 5.5 The Petzval lens file has been modified such that the lens now corresponds with the 143-mm Petzval lens we worked with in the last chapter. The text window shows new lens data, while the second graphics window shows aberration curves (ray-trace analysis).

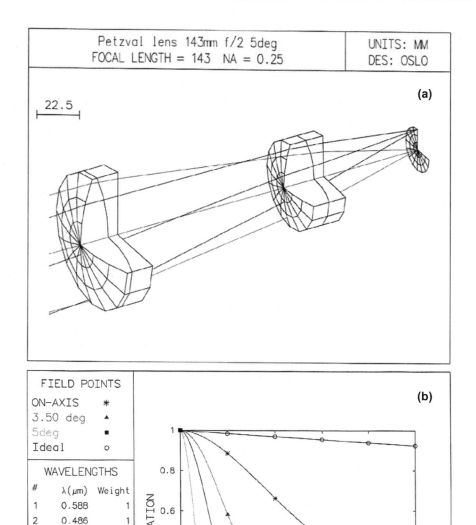

Figure 5.6 Following the basic design procedures, numerous evaluation routines are available in OSLO-EDU. Shown here are a solid model lens drawing (a) and MTF curves (b) that permit estimates of lens resolution.

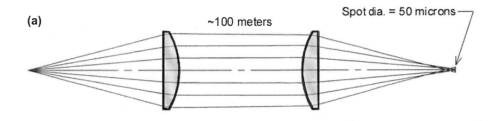

(a)

~100 meters

Spot dia. = 50 microns

Simple plano-convex lens configuration

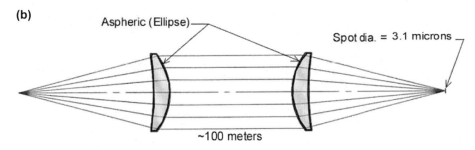

(b)

Aspheric (Ellipse)

Spot dia. = 3.1 microns

~100 meters

Optimized meniscus lens configuration,
with convex aspheric surface

Figure 5.7 Optical system used to transmit laser energy from point to point. The initial simple lens system (a) has excessive spherical aberration. Residual aberrations are corrected by modifying the lens form to a meniscus shape with an aspheric surface (b).

a monochromatic system with no extended field, spherical aberration will be the only image quality concern. We want to keep the two lenses identical, thus there are only two ways that the residual spherical aberration can be reduced. They are: modify the lens shape or make one of the lens surfaces aspheric. Using the very fundamental optimization routines offered in OSLO-EDU, it is possible to generate the optimized design shown in Fig. 5.7(b). In this optimized solution the lens shape has been made a meniscus and the convex surface is an ellipsoid rather than a sphere. Knowing this to be an application where many hundreds of these lenses will be needed, making one surface aspheric will not be a problem, assuming modern methods of lens machining or lens molding will be implemented. Analysis of image quality for a lens pair indicates the spot formed at the image surface of the second lens will be a diffraction-limited Airy disk with a diameter of 3.1 μm. This is compatible with the 5-μm fiber diameter that will be located at that point. The best method of image evaluation in this case would be to use the radial energy distribution subroutine that is part of the point-spread function evaluation. Figure 5.8 shows the point-spread function data, including the radial energy distribution (lower-right). These calculations are based on a combination of geometric and diffraction effects.

Using OSLO-EDU, it is possible to examine the effects of manufacturing and assembly tolerances. For example, we will consider the impact of decentering the projection lens element relative to the input fiber optic bundle (see Fig. 5.9). It is reasonable to assume that, due to the machining set up, the fiber optic bundle and the bore that will hold the lens will be essentially concentric. On the other hand, to facilitate assembly, it is essential that the lens diameter be less than the bore diameter in all cases. For the dimensions shown, with a nominal lens diameter of 2.440 in. and a nominal bore diameter of 2.442 in., there will be a nominal 0.001-in. clearance around the lens. In the worst case that clearance will increase to 0.002 in. If the lens is shifted, i.e., decentered by 0.002 in., the transmitted wavefront error will increase from 0 to 0.51λ. In this application, that would be judged a statistically acceptable "worst case" error.

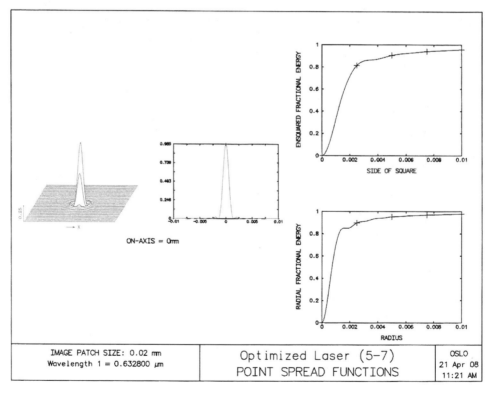

Figure 5.8 Point-spread-function data for the optimized laser lens system. Radial energy curve (lower-right) shows 85% of the energy to be within a 1.8-µm radius.

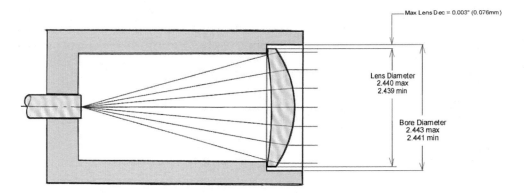

Figure 5.9 Tolerance analysis showing lens centration relative to the input fiber optic bundle.

5.6 Instrument Design and Analysis

This exercise will incorporate the thin-lens concepts introduced in the last chapter, combined with the computer-aided system layout, catalog lens selection, and system-analysis features of the OSLO software package, to design a complete optical system.

This design will involve a low-power microscope with the following specifications:

Magnification .... 10×
Object diameter .... 12.0 mm
Working distance .... >90 mm
Exit pupil diameter .... 3.0 mm
Eye relief .... >20.0 mm
Image quality .... Eye limited

A microscope generally consists of two lens assemblies: the objective lens and the eyepiece. Starting with the two-element thin-lens layout shown in Fig. 5.10(a), one reasonable approach would be to assume that the objective lens will operate at a magnification of 1× and the eyepiece at 10× (EFL = 25 mm). The 90-mm working distance dictates an objective lens focal length of about 50 mm, while the overall distance from object to image will be approximately 200 mm.

Now, considering the actual configuration of the final system, the thin-lens arrangement can be modified such that it is made up of commercially available achromatic doublets that will be found in vendor catalogs. A variety of such lenses are a part of lens libraries found in OSLO and most other optical design software packages. Figure 5.10(b) shows the real lens system in a form that consists of four cemented achromats, two each of two identical catalog lenses. The eyepiece focal length will be about 25 mm, making it a 10× magnifier. The

(a)

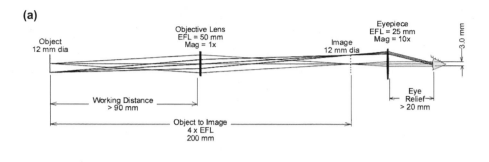

(b)

Figure 5.10 The thin-lens layout of a 10× microscope (a) consists of an objective lens and an eyepiece. This is converted to a real design (b) using acromatic doublets selected from vendor catalogs found in the OSLO software package.

3-mm exit-pupil diameter leads to an eyepiece $f/\#$ of $25/3 = f/8.3$. For the 1× objective, both the input and output $f/\#$ will also be $f/8.3$. Thus, the height of the marginal ray at the thin-lens objective will be about $(100/8.3)/2 = 6$ mm. Thus, our 1× objective will need to have a clear aperture diameter of about 12 mm.

Also, from the thin-lens layout, it is possible to estimate the eyepiece field of view. The image size at the eyepiece focal plane is equal to the object size, 12 mm in diameter. The half field of the 25-mm eyepiece then will be the angle whose tangent is equal to 6/25, which is 13.5 deg. With this preliminary thin-lens system layout and analysis done, it is now possible to use the computer with the OSLO software to select a set of real catalog optics and to evaluate the detailed optical performance of the system. Unfortunately, because this will be a system containing four cemented achromats (12 surfaces), an object and image surface, an internal image surface and an exit pupil surface, we will obviously exceed the ten-surface limit of the OSLO-EDU program. The remainder of this example has been generated using one of the other available versions of OSLO.

The first step in building the lens system shown in Fig. 5.10(b) is to select a pair of identical doublets with a focal length of 100 mm and a diameter greater than 12 mm. These two lenses are arranged as shown and adjusted axially relative to the object surface to produce a 1:1 object-image relationship. In order to clearly define the field being viewed, a 12-mm-diameter mechanical field stop will be placed at the internal image surface as shown. Next, a pair of 50-mm-

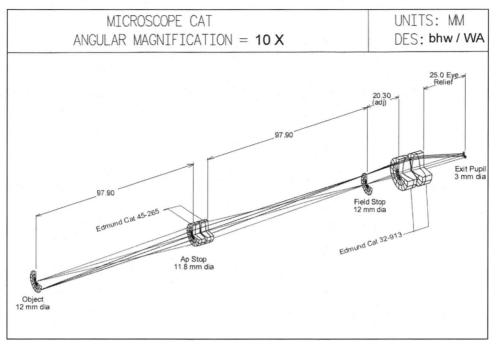

Figure 5.11 10× simple microscope, designed using OSLO computerized software package. All lenses have been selected from available catalog stock as listed in the software lens library.

focal-length doublets with a diameter of 18 mm are selected and arranged as shown to make up the 25-mm (10×) eyepiece. The space from the internal image to the eyepiece is adjusted to make the output from the eyepiece collimated, i.e., focused at infinity. Ray tracing of the off-axis bundle reveals that the eye relief (distance from the last lens vertex to the exit pupil) of this system will be 25 mm. This determination is based on the positioning of the system aperture stop (entrance pupil) at the first surface of the objective lens. This aperture stop would be a mechanical aperture with a clear diameter of 11.2 mm. This value can be conveniently determined by actual ray tracing of the system.

Most lens systems produce an output that is focused to an image surface where image analysis takes place. Such a system is referred to as a "focal" system when it is set up for computer analysis. In a case such as this microscope system, the output is collimated; thus, the image surface is located at infinity. This type of system is referred to as an "afocal" system, and the residual aberrations will be presented as angular, rather than linear values. Figure 5.11 shows details of the final microscope system as it would be constructed using achromatic doublets selected from the Edmund Scientific catalog. Figure 5.12 shows the ray-trace data and the MTF curves for this design. MTF data shows clearly that, for the on-axis case, the optical system is diffraction limited. The aerial image modulation (AIM) curve for the eye is superimposed on the MTF curves to show the actual visual resolution that will be achieved when viewing

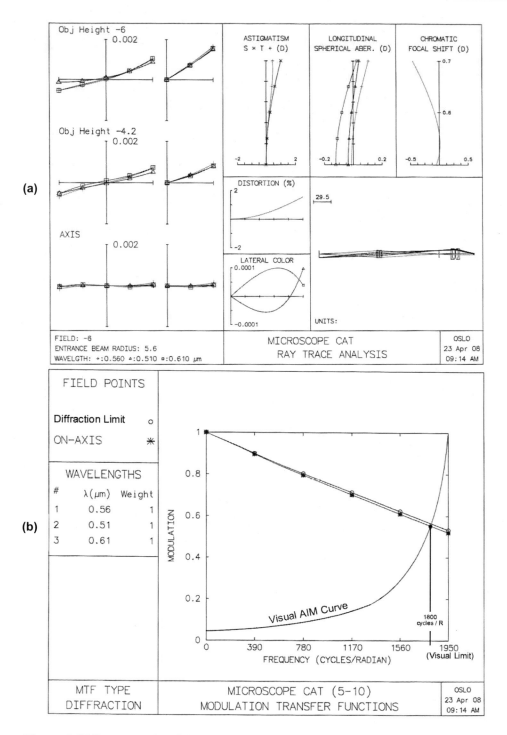

Figure 5.12 Ray-trace data (a) and MTF curves (b) for a 10× microscope designed using Edmund catalog optics from the OSLO lens library.

through this microscope. For the on-axis case, the data shows that the visual resolution limit (1950 cycles/rad) will be degraded by the MTF of the optics (primarily due to diffraction) by about 8% to a value of 1800 cycles/rad. Glancing at the aberration curves, it can be seen that the major residual aberration is field curvature. Since field curvature and astigmatism are less than one diopter, it can be safely concluded that resolution over the full field of view will be satisfactory. Aberration data also shows distortion to be less than 2%, while both axial and lateral (off-axis) color are well corrected. This all represents good optical performance for a system made up using about $250 worth of readily available catalog optics.

A few interesting points may be revealed by changing the approach used to analyze this optical system. The entire lens configuration can be reversed, which will permit ray tracing from an object at infinity through the 3-mm-diameter exit pupil, through all of the optics, and then onto a 12-mm-diameter image surface (was the object). This will require that we change the computer evaluation mode from afocal to focal. The optical system will now appear as shown in Fig. 5.13. On-axis MTF data has been calculated with the result shown in Fig. 5.13. When used with a 10× microscope, the typical eye (which has a 3-mm pupil) will be able to resolve 77 cycles/mm at the object. As we found in our afocal evaluation, the MTF curves here indicate that the visual resolution will be reduced by about 8% to 71 cycles/mm, most of which is due to diffraction effects. This design will meet the general specification that the resolution capability of the 10× microscope be essentially eye limited.

While this design represents a relatively economical solution ($250 for the lenses), the designer will almost always be faced with the question, *How can we reduce cost?* With a computer model to use for evaluation, it is fairly simple to explore cost-saving approaches. For example, we might consider replacing the doublet nearest the eye with a less-expensive plano-convex singlet lens element. Examination of the resulting aberration curves indicate that the amounts of residual field curvature and lateral color will be increased to 1.5 diopters and 5 min, respectively. These increases do not seem to be justified by a savings of about $30. Of course, the final decision in these matters will always be made by the customer, with knowledgeable advice and inputs from the optical designer.

5.7 Magnification Analysis

This section will deal with the topic of magnification in an optical system. In the case of the microscope that was just designed, the initial specification called for a magnification of 10×. In a basic compound microscope the magnification is found by multiplying the magnification of the objective lens by that of the eyepiece. In this design we assumed a magnification of 1× for the objective and 10× for the eyepiece. Let's now examine the final design and determine the actual magnification of that system. For a visual system such as this, the magnification is generally defined as a comparison of the object's apparent size when it is viewed with and without the optical device.

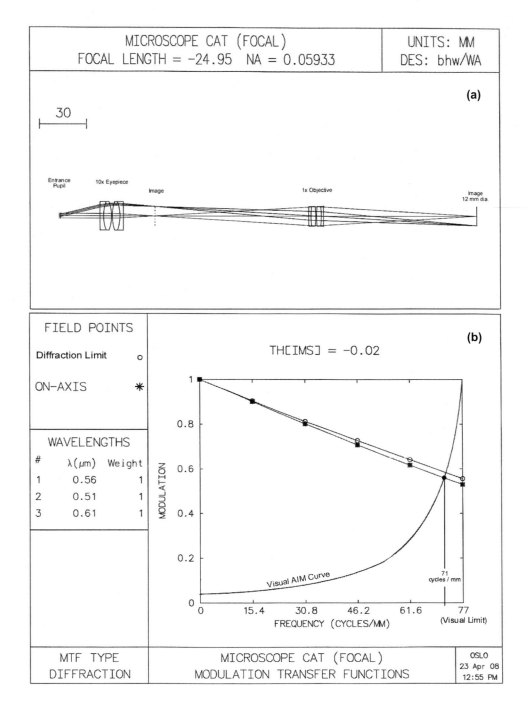

Figure 5.13 Ray-trace data (a) and MTF curves (b) for a 10× microscope evaluated in the "FOCAL" mode.

In this case the object is a 12-mm-diameter circle. When viewing with the naked eye, in order to resolve maximum detail, the viewer will bring the object to the closest point of clear focus for the typical eye. This standard *near point* of vision has been established to be 10 in. (254 mm). At that distance from the eye, the 6-mm object height will subtend the angle whose tangent is 6/254 = 0.0236. Assuming an eye pupil diameter of 3 mm, a viewer with normal vision will resolve 7.7 cycles/mm at this object distance. This corresponds to resolution of 65 μm at the object.

In the computer model of the final microscope design shown in Fig. 5.10, the eyepiece focus can be adjusted so that the final image is formed at a point 254 mm to the left of the exit pupil (viewer's eye location). The resulting configuration is shown in Fig. 5.14. The basic magnification is determined by dividing the object height into the image height. In this case the result will be 60.5/6.0 = 10.1×. The 1% difference between our specification and our final result is negligible. Its source can be attributed to residual distortion and/or the fact that we selected a 25-mm focal length for our 10× eyepiece, while a value of 25.4 mm would have been more precise. If the magnification requirement was more critical (which is not likely), the final magnification could be fine tuned by adjusting the air space between the two catalog lenses that make up the eyepiece.

In terms of visual resolution, as we said earlier, at a distance of 254 mm the typical eye will resolve 7.7 cycles/mm, or 65 μm at the object. As our earlier MTF analysis showed, when viewing through the microscope, this visual resolution will be reduced by about 7%, to 7.1 cycles/mm, or 70-μm elements at the final image. You will recall that this reduction in resolution is due primarily

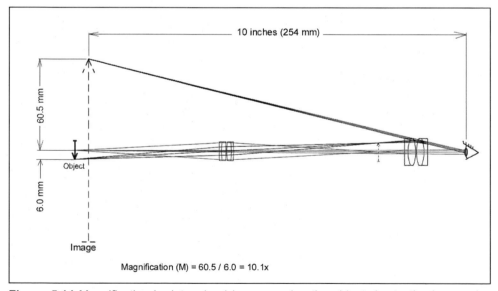

Figure 5.14 Magnification is determined by comparing the object size to the image size when focus is set to form the final image 254 mm from the viewer's eye.

to diffraction effects of the microscope optics. Nonetheless, the advantage of magnification remains in that the object now appears 10 times larger to the viewer's eye. As a result, the number of resolved elements across the 12-mm-diameter object will increase from 12.0/0.065 = 185 elements, to 121/0.070 = 1729 elements. Thus, we can conclude that viewing the object with our 10× microscope will result in a 1729/185 = 9.35× net increase in our ability to resolve detail in that object.

5.8 Design of a Noncatalog System

While the ability to design systems with lenses from the software catalog base is a very useful function, there are many other optical design tasks that can be accomplished outside of that realm. This example will demonstrate one of those tasks. Quite often when reviewing new material in optics-related literature, we are inspired to generate a new system design that incorporates this material. For example, suppose we find data that describes a new form of doublet lens that offers significant advantages over more traditional forms. This might lead us to thinking about the design of an *ultimate* refracting telescope, suitable for use in the field of serious amateur astronomy. Using a modern optical design software package it will be possible to generate just such a design.

The literature often contains an illustration and a table of lens data such as the one shown in Fig. 5.15. This is an 800-mm, *f*/10 semiapochromat that would make a very nice objective lens for a compact refracting telescope. The correction of chromatic aberration (color) in a telescope objective is always a major goal. In an *achromat*, the lens form is optimized such that the red and blue light are brought to a common focus. The difference between this common focus and the focus for green light is referred to as secondary color. This secondary color is the residual aberration that limits the on-axis image quality of a conventional, well-designed achromatic doublet. The spot size for a well-corrected, 800-mm, *f*/10 achromat will be about 25 μm in diameter, due primarily to residual secondary color. The term *apochromat* is used to designate a design where the red, green, and blue light are all brought to a common focus. This level of correction will lead to diffraction-limited image quality. The spot size for a typical diffraction-limited *f*/10 apochromat will be 13 μm in diameter, essentially doubling the image quality of a conventional achromat. An apochromatic objective lens will generally require three or more lens elements that include the use of very exotic (read: expensive and difficult to work) optical glass types. The major advantage claimed for this new, semiapochromatic design is diffraction-limited performance over the visual spectral bandwidth, accomplished using just two elements, neither of which is terribly exotic. Pound for pound, the glass used for element II in Fig. 5.15 will cost about 10× the cost of conventional crown glass, and it will be a bit more difficult to work in the optical shop. Neither factor is a showstopper, so we will proceed with the design and analysis of a telescope incorporating this objective lens.

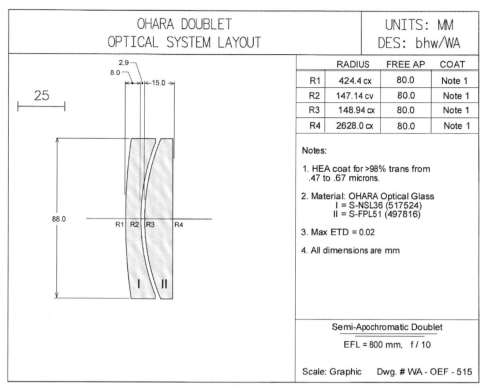

Figure 5.15 Typical drawing showing design data on a new semiapochromatic air-spaced doublet lens design.

First, it is important that we confirm the validity of the claims for this new design. Quite often the designer is reticent to divulge all details of a new design. To check this design we will create an OSLO lens file based on the information contained in Fig. 5.15. The claim for diffraction-limited performance can be confirmed by calculating the MTF data for the lens design. This has been done, with the results shown in Fig. 5.16. This plot clearly shows that the on-axis MTF curve is essentially coincident with the diffraction limit (ideal) and that the off-axis performance is also very well corrected. Knowing this, it is now reasonable to proceed with our telescope design.

A lens layout will allow us to establish the basic optical characteristics of the telescope. The objective lens will take the form shown in Fig. 5.15. It will have an aperture of 80 mm, a focal length of 800 mm, and a half-field angle of about 0.8 deg. We will assume the use of a 28-mm focal length RKE® eyepiece from the catalog of Edmund Optics. Prescription data for this eyepiece is available from Edmund, and it is also available in lens libraries of several software packages. This eyepiece data is added to the objective lens to make up the complete telescope as shown in Fig. 5.17. Telescope magnification M is calculated by dividing the objective lens EFL by the eyepiece EFL, yielding a result of $M = 800/28 = 28.5 \times$. Catalog information tells us that the eyepiece

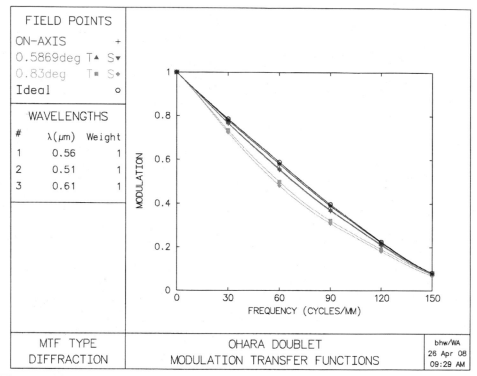

Figure 5.16 To confirm the validity of the doublet design, lens data is entered into OSLO and MTF data is calculated. The above result indicates true diffraction-limited image quality.

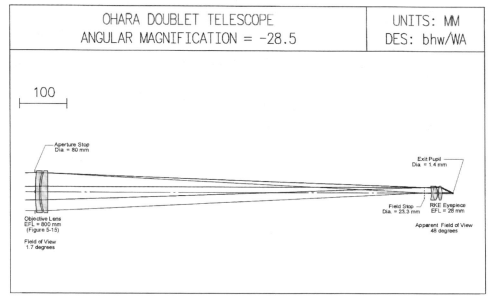

Figure 5.17 Design layout of a 28.5× telescope made using the doublet from Fig. 5.15, combined with a 28-mm RKE® eyepiece from the Edmund Optics catalog.

assembly contains a 23.3-mm-diameter field stop. The half-field angle of the telescope is adjusted to 0.834 deg to agree with this value. With the system focus set for collimated output, and the evaluation mode set to afocal, the ray-trace data shown in Fig. 5.18 is generated. From this data we can conclude that, in addition to well-corrected, on-axis aberrations, lateral color is also well corrected and that residual field curvature and astigmatism will be about 1 D. The residual distortion of 17% should not be problematic for normal astronomical viewing. If reduced distortion is desired, that can be easily accomplished by substituting an eyepiece with a more sophisticated design form. Also of interest, the 28-mm RKE[®] eyepiece is one of a set of available eyepieces with focal lengths of 8, 12, 15, 21, and 28 mm. In this telescope, these eyepieces will result in magnifications of 100×, 67×, 53×, 38×, and 28.5×, respectively. While all of these eyepieces would not be necessary, at about \$60 each it would seem reasonable to have a set containing three or four of these eyepieces.

To summarize the performance of this telescope, we can consider the MTF data shown in Fig. 5.19. The curves are plotted out to the visual limit of 1780 cycles/rad. This is the angular resolution limit for the typical eye with a pupil size of 1.4-mm diameter. The AIM curve for the eye is plotted and the resolution of the combined telescope plus eye is found to be 1660 cycles/rad. As we saw earlier in the microscope example, this 7% reduction in visual resolution is due primarily to unavoidable diffraction effects of the telescope optics.

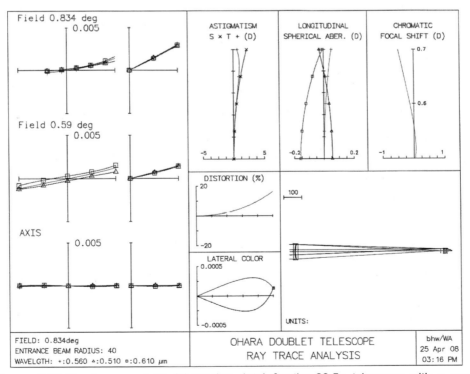

Figure 5.18 Ray-trace analysis (aberration data) for the 28.5× telescope with a new, semiapochromatic objective lens and a commercially available RKE[®] eyepiece.

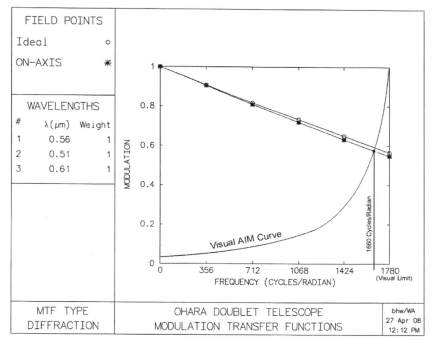

Figure 5.19 MTF data for complete 28.5× telescope with semiapochromatic objective (Fig. 5.15), and 28-mm RKE® eyepiece.

In more meaningful terms, assume we are viewing the surface of the moon on a clear night. If our eye pupil is fixed at 1.4-mm diameter, we will resolve detail down to 1780 cycles/rad. This corresponds to angular resolution of 116 arcsec per cycle. This means that, with the naked eye, we could resolve two distinct small craters on the surface of the moon that were separated by a distance of 126 miles. If they were any closer together, they would not be resolved as two separate craters. Viewing through our 28.5× telescope, our resolution capability would be increased to 1660 × 28.5 = 47319 cycles/rad. This corresponds to angular resolution of 4.36 arcsec per resolved cycle. This means that, through the telescope, we could now resolve two distinct small craters that were separated by a distance of just 4.8 miles. As a result, the net gain in resolving power with versus without the telescope is found to be $126/4.8 = 26.25\times$.

5.9 Review and Summary

This chapter has dealt primarily with the application of today's readily available personal computers and computer software packages to the design and analysis of a variety of optical systems. In order to demonstrate the significance of these modern developments, a historic perspective has been presented dealing with the field of lens design in general and the development of the OSLO software package in particular. One of the major points that evolved from this historic review has been the shifting of engineering responsibilities that has taken place in recent years. This has resulted in a significant overlap in the responsibilities of the optical engineer with those of the lens designer. Although *optical designer* is a familiar title, one that has existed for years, this chapter has attempted to define the work of the optical designer in terms that are consistent with today's workplace conditions.

Several lens design examples have been presented with the aim of illustrating the broad scope of work that can be handled, the flexibility of these software packages, and the ease with which they can be used. Perhaps the most significant fact is that these programs can and will be used by a broad cross section of those involved in the science of optics and optical engineering. What was once the exclusive domain of the lens designer has become an area that can and should be explored by many who are working in the field of optics. At the same time, this new computerized approach to optical design allows the lens designer to expand his work into areas that were not previously accessible. The net result of all this will be increased productivity and better solutions to a wide range of optical design problems. It also makes the work of all involved in the field of optical engineering more challenging and more exciting.

Chapter 6
Primary Lens Aberrations

6.1 Introduction

It is the function of a lens to collect light from a point on the object and to focus that light to a corresponding point (a conjugate point) on the image. In nearly every case the lens will fail at this task, in that there will be some residual error in the precision with which the lens collects, refracts, and focuses that light. Rather than a true point image, the lens will produce a blur circle, i.e., a spot. It is the function of the lens designer to ensure that this spot size is sufficiently small to allow the lens to produce the required resolution, or image quality.

These errors in the lens's ability to form a perfect image are referred to as *lens aberrations*. There are seven primary aberrations that must be considered when a lens system is being designed or evaluated. By understanding the basic characteristics of these seven primary aberrations, the optical engineer will be better equipped to specify and evaluate the image quality of a lens or an optical system. The following paragraphs will describe the seven primary aberrations and discuss the more important characteristics of each.

6.2 Spot Diagram, Radial Energy Distribution, and Modulation Transfer Function

There are several methods of computer analysis available to assist in the visualization and evaluation of the image quality that will be produced by a lens or an optical system. For purposes of this discussion of primary lens aberrations, we will present three of the more commonly used methods: the spot diagram, radial energy distribution, and MTF.

A spot diagram is generated by tracing a large number of light rays (typically 200 to 300) from a point on the object, through the lens system, and then onto the image surface of that lens. The spot diagram is created by plotting the intersection point for each ray on the image surface. This spot diagram represents a reasonably accurate simulation of the image of a point-source object (such as a star) as it will be produced by the lens. If the lens were perfect, all rays would intersect the image at the same point and the spot diagram would be just that, a point. This purely geometric ray-trace analysis neglects the effects of diffraction,

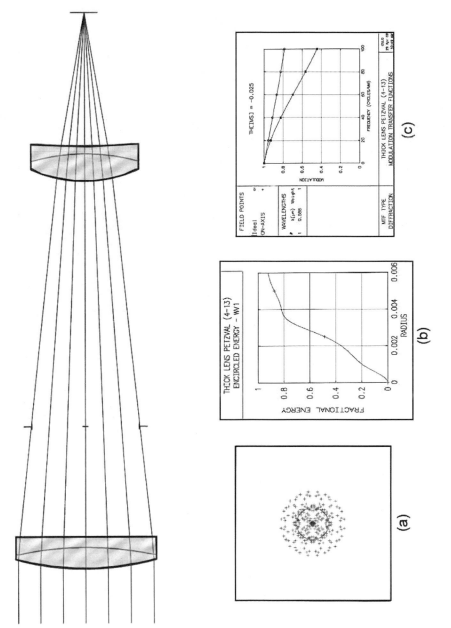

Figure 6.1 Well-corrected Petzval lens (top). Analysis showing (a) spot diagram, (b) radial energy distribution, and (c) MTF data.

which were discussed earlier (see Fig. 3.14). For any lens that approaches being diffraction limited (wavefront errors <1λ), it is essential that diffraction effects be included in the analysis of its image quality.

For example, the Petzval lens shown at the top of Fig. 6.1 has residual on-axis aberrations that produce an on-axis wavefront error of about ½λ. The spot diagram contains an embedded reference circle which represents the size of the Airy disk for this lens. In this case, diffraction effects will affect subsequent image analysis. The radial energy distribution (RED) curve shown in Fig. 6.1(b) is a result of combining both geometric and diffraction effects. Likewise, the MTF curves indicate that, if the lens were diffraction limited, the modulation (contrast) of the image at a frequency of 100 cycles/mm would be nearly 80%. Because of the residual lens aberrations, that image modulation will be reduced to about 45%.

6.3 Spherical Aberration

There are three basic types of lens aberrations: on axis, off axis, and chromatic. The principal on-axis aberration is spherical aberration. This is the imaging error found when a lens is called upon to focus an axial bundle of monochromatic (single wavelength) light to a point image. In the presence of spherical aberration, each zone, or annulus, of the lens aperture will be found to have a slightly different image distance. The result can be seen in Fig. 6.2, which illustrates the presence of spherical aberration in a simple biconvex lens. In this example, spot diagrams have been generated at three separate focus positions. Focus point (1) represents the paraxial focus. At this point those rays very close to the axis will be focused. However, rays passing through zones farther away from the lens center will be focused progressively short of the paraxial focus due to spherical aberration. The farther the rays are from the lens center, the greater the error in focus. This lack of a common focus, or image distance, for all zones of the lens aperture is, by definition, *spherical aberration*.

It can be seen from Fig. 6.2 that there is a point short of the paraxial focus where the blur circle, or spot size, due to spherical aberration is at a minimum [Fig. 6.2, point (2)]. Point (2), which is midway between points (1) and (3), has also been selected for analysis. Figure 6.2 shows spot diagrams representing the spot makeup at points (1), (2), and (3). A good sense of the image quality of this lens in the vicinity of the paraxial focus can be gained by a careful study of these spot diagrams.

Figure 6.3 shows the RED curves that correspond to the three spot diagrams shown in Fig. 6.2. Analysis of these RED curves will yield some indication of the relative image quality at each point of focus. In general, the smaller the spot, the better the image quality. Reviewing the three RED curves, it appears at first glance that the focus position for minimum spot size (3) would represent the point of best image quality. However, it can be seen that at the 25% fractional energy level, the spot radius at the middle focus position (2) is about half the size of the corresponding spots for the other two focus positions. Conversely, at about

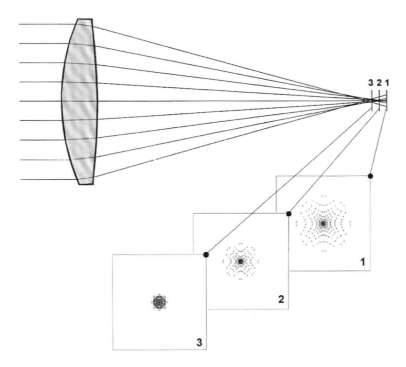

Figure 6.2 Simple lens with spherical aberration. Spot diagrams are shown for points of paraxial focus (1), minimum spot size (3), and midway between the two (2).

a 65% fractional energy level we see that the paraxial focus RED curve (1) is about twice the size of the corresponding spots for the other two focus positions. One reasonable way to choose the RED curve that is representative of the best image quality is to select the curve with the minimum area to its left. In nearly all applications where spherical aberration is the dominant residual aberration, overall image quality will be best when focus is set to a point close to the point of minimum spot size (3).

Figure 6.3 also contains the MTF curves for this simple lens at the established focus points 1, 2, and 3. MTF calculations assume an object that is made up of a series of high-contrast black and white parallel bars whose frequency varies from zero to the maximum frequency of interest. The MTF curves indicate the fall off in image modulation (contrast) as the frequency of the black and white bars at the image increases. Considering MTF curve 3, it can be seen that at lower frequencies (0 to 1.5 cycles/mm), the contrast of the image will exceed the contrast found at the other two focus positions. Unless the higher frequencies were of some particular importance, it would once again be concluded that focus position 3, the point where minimum spot size occurs, would be the point of best focus. In addition to determining the point of best focus, this brief analysis also gives a good feel for the sensitivity (in terms of image quality versus focus), for this particular lens.

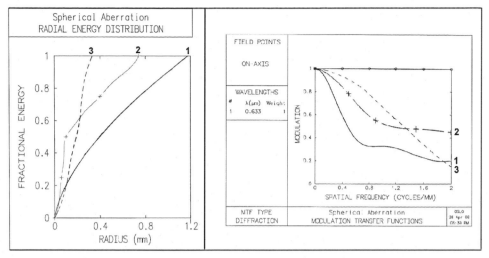

Figure 6.3 Radial energy distribution (RED) and modulation transfer function (MTF) curves, through focus, for a simple lens with spherical aberration (Fig. 6.2).

In the case of a simple lens element, the number of tools available to control spherical aberration are limited. First, while maintaining the power (EFL) of the lens constant, the lens shape may be adjusted such that the amount of residual spherical aberration is minimized. This action is referred to as bending the lens. A second approach to reducing spherical aberration would involve increasing the index of refraction of the lens material. If the optical glass of the lens shown in Fig. 6.2 is changed such that the index increases from 1.51 to 1.91 and the lens bending is modified to minimize residual spherical aberration, the spot diameter at the point of best focus will be reduced from 0.50 mm down to 0.19 mm, about a 62% reduction. As always, trade-offs will be involved. In this case the high-index glass will be more expensive, it will be more difficult to work in the optical shop, and it will introduce more residual chromatic aberrations, i.e., color. Note that any residual spherical aberration will have an equally negative effect on off-axis image points.

While the amount of residual spherical aberration can be minimized, it cannot be totally eliminated when a real object and image are involved. For this reason it is essential that the optical designer establish just how much residual spherical aberration will be allowable, while still producing an acceptable lens design. In order to make this determination, the optical designer must develop a sense of just how this optical system will function and what levels of image quality will be consistent with other system components (such as the detector) and with the overall system specifications.

6.4 Coma

Coma is an aberration that affects off-axis light bundles in a manner that is similar to the way that spherical aberration affects on-axis light bundles. In the case of coma, the fact that the light bundle passes through the lens at an angle results in a total lack of symmetry in the aberrated spot. Rather than a circular spot, a lens that is afflicted with coma will generally produce a more or less triangular spot for off-axis image points (see Fig. 6.4).

Figure 6.5 shows a spot diagram for an image point where coma is the predominant aberration. Also shown is the corresponding RED curve for this spot. Because the comatic spot is not symmetrical, the centroid for the RED calculation must be selected statistically.

For a pair of simple lenses arranged symmetrically about the system aperture stop, as in a 1:1 relay lens or a complex lens that has some degree of symmetry about its aperture stop, there will be a significant reduction in the amount of residual coma found. This is the case because the coma that is introduced when passing through the first half of the lens will be partially or completely cancelled out when passing through the second half. This important characteristic is used in the design of many lenses and optical instruments, such as the double-Gauss lens, borescopes, and submarine periscopes. However, the residual coma in a lens system will usually be accompanied by some other off-axis aberrations, such as astigmatism, making its individual contribution to final image quality difficult to evaluate.

6.5 Field Curvature

In most optical systems it is required that the final image be formed on a flat surface, i.e., the image plane. Unfortunately, it is the natural tendency of most systems to form that final image on a curved surface. The nominal curvature (1/radius) of that preferred image surface is referred to as the Petzval curvature, or *field curvature,* of the system. For a simple lens, this field curvature will be approximately equal to two-thirds of the lens power (1/EFL). In other words, for

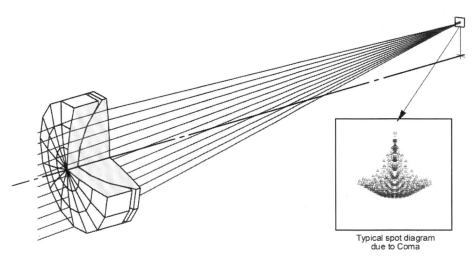

Typical spot diagram
due to Coma

Figure 6.4 Coma is an off-axis aberration that produces an asymmetrical spot at the image surface.

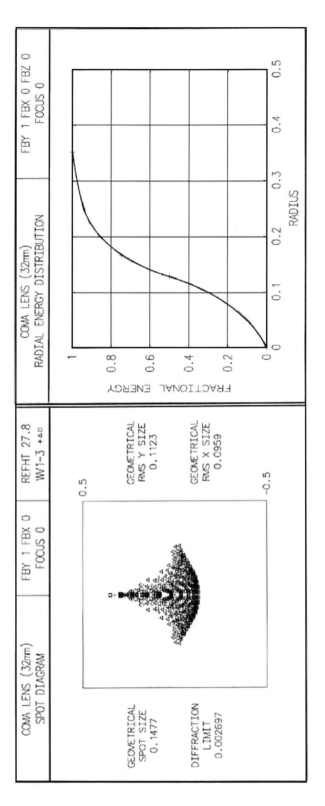

Figure 6.5 Spot diagram and RED curve for an image afflicted with coma.

a simple lens, the radius of the preferred image surface will be equal to about 1.5× the EFL of that lens. When the lens is free of all other aberrations, an essentially perfect image will be formed on that curved (Petzval) surface. When residual astigmatism is present, as is almost always the case, the Petzval surface has little real significance as far as the actual image quality of the lens system is concerned.

6.6 Astigmatism

When astigmatism is present in a lens system, off-axis ray fans that are oriented tangentially and sagittally at the lens aperture will focus onto two distinctly separate curved surfaces. Figure 6.6 shows two such ray fans—one in the tangential (Y-Z) plane, the other in the sagittal (X-Z) plane—passing through a simple lens. The figure shows how these ray fans will be focused by the lens when astigmatism is present. It can be seen here, and also in the corresponding sag curves and spot diagram shown in Fig. 6.7, that the presence of astigmatism will cause the ideal circular spot to be blurred into an elliptical, or astigmatic shape. Since this elliptical spot has significant differences between its horizontal and vertical dimensions, it follows that, when astigmatism is present, the resolution will vary depending on the orientation of the line pattern in the object being imaged. Because of the extreme lack of symmetry in the spot, the RED calculation is not meaningful and will seldom be performed on an astigmatic spot.

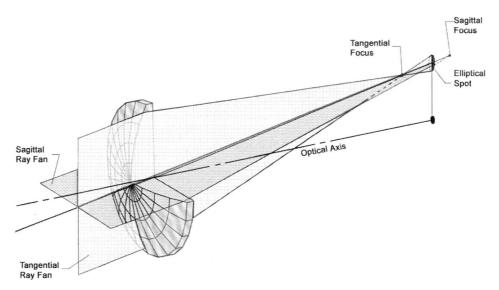

Figure 6.6 Astigmatism is the off-axis lens aberration caused by the lens having a different focus for the tangential and sagittal off-axis ray fans. The result will most often be an elliptical rather then circular spot image of a point source.

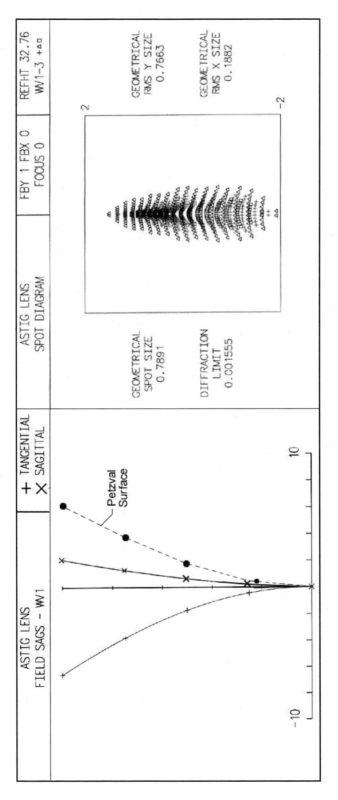

Figure 6.7 Field sag curves anc spot diagram for an image with astigmatism.

The field sag curves shown in Fig. 6.7 represent the most common method of illustrating and evaluating the aberrations of field curvature and astigmatism. These curves represent a cross section through half of the image surface, from the optical axis out to the edge of the field. In the case shown here, the system focus has been set so that the tangential fan focuses short of the chosen image surface, while the sagittal fan focuses beyond it (as was shown in Fig. 6.6). If we consider the case where the image being formed is a spoked wheel, the rim of the wheel will be in focus at the top of the tangential curve, while the spokes will be in focus along the sagittal curve. Astigmatism is, by definition, the difference between the tangential and sagittal curves. If the tangential and sagittal curves are coincident, then the lens is said to be free of astigmatism and the resulting curved image will be formed on the Petzval surface. When astigmatism is present, the distance from the Petzval surface to the tangential field curve will be three times the distance from the Petzval surface to the sagittal field curve (see Fig. 6.7). In most cases it is not possible to correct both field curvature and astigmatism to zero, but satisfactory image quality can be often achieved by balancing residual astigmatism with inherent field curvature.

Another effective approach to the correction of field curvature and astigmatism is to first correct the lens design such that there is little or no residual astigmatism and the image is formed on the Petzval surface. Once this is accomplished, a field flattener lens can be added to the design in close proximity the final image. The Petzval curvature of the field flattener is made equal to and opposite in sign to the Petzval curvature of the basic lens. The net result will be a lens design that is free of both astigmatism and field curvature. The drawback to this design approach (remember, there are always trade-offs) lies in the fact that the field flattener lens must be located very close to the image plane. In the worst case, some mechanical interference condition may exist, making this approach physically impossible. In another case we will find that the close proximity of the lens to the image will mean that the lens surfaces must be meticulously clean, free from scratches, dirt, and dust which, if present, would appear slightly out of focus, superimposed on the final image.

6.7 Distortion

Distortion is a unique aberration in that it does not affect the quality of the image in the usual terms of sharpness and focus. Rather, distortion affects the overall shape of the image, causing it to depart from a true-scaled replica of the object. Figure 6.8 shows three lenses. The first (a) is totally free of distortion and, as a result, it produces a true reproduction of the "checkerboard" object. If the lens is afflicted with positive distortion, then the off-axis points will be imaged at distances from the axis that are greater than nominal, resulting in the pincushion effect that is seen in Fig. 6.8(b). On the other hand, if the lens is afflicted with negative distortion, then the off-axis points will be imaged at distances from the axis that are less than nominal, resulting in the barrel shape that is seen in Fig.

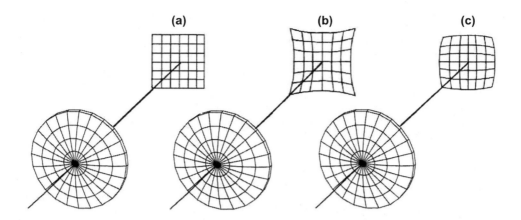

Figure 6.8 Distortion is present when the focal length of the lens varies as a function of field angle or image size. In this illustration, (a) the lens on the left is free of distortion, (b) the lens in the center has 16% positive (pincushion) distortion, and (c) the lens on the right has about 10% negative (barrel) distortion.

6.8(c). For most visual systems distortion errors of about 5 to 7% will usually be considered acceptable. For camera and projection lenses distortion values of 1 to 2% will generally represent the upper limit of acceptable.

Note: The five aberrations presented to this point have been monochromatic aberrations, generally calculated at the central wavelength of the lens system. If the lens is to be used over an extended spectral bandwidth, the following chromatic aberrations must be taken into account.

6.8 Axial Color

In Sec. 3.9, which dealt with dispersion, it was stated that the index of refraction for all optical glasses will vary as a function of wavelength. The index will be greater for the shorter wavelengths (blue), and the rate at which the index changes will also be greater at these shorter wavelengths. In a simple lens this results in each wavelength being focused at a different point along the optical axis. While this chromatic spreading of white light by a prism is referred to as dispersion, when it occurs in a lens, as described here, it is referred to as *primary axial color*. Figure 6.9(a) illustrates a simple lens focusing an axial bundle of white light. If the focus is set for the middle of the visual spectral band as shown, the blur circle will consist of a yellow-green central core with a halo of purple (red plus blue) light surrounding it. Except in very unique cases, such as laser systems or other nearly monochromatic systems, axial color is an aberration that must be dealt with in order to achieve reasonable image quality. This is most easily accomplished by converting the simple lens into an achromatic doublet as

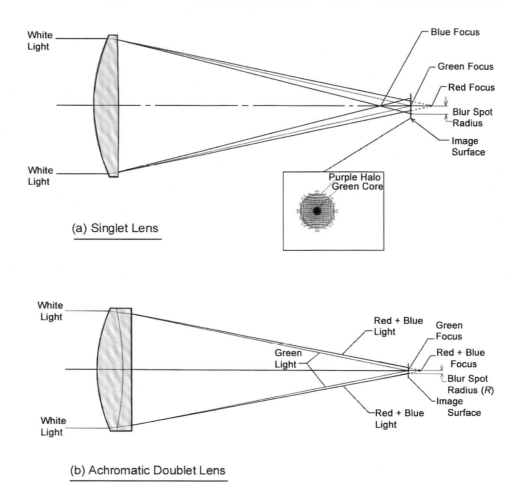

(a) Singlet Lens

(b) Achromatic Doublet Lens

Figure 6.9 When an achromatic doublet is substituted for a simple lens, the blur-spot radius R due to chromatic aberration will be reduced by about 10×.

shown in Fig. 6.9(b). Here, two glass types and lens powers are selected such that the primary axial color has been corrected by bringing the two extreme wavelengths (red and blue) to a common focus. In most cases the substantial reduction in blur-circle size due to achromatization (about 10× for the visual spectrum) will result in a satisfactory design.

When this state of achromatization is achieved, the *primary color* is said to be corrected. The remaining chromatic aberration is referred to as *secondary color*. As Fig. 6.9(b) shows, this is due to the difference in focus between the extreme wavelengths (red and blue) and the central wavelength (green). Secondary color is often found to be the limiting axial aberration in a lens design that has otherwise been carefully designed and optimized. Because of this, it is often useful to be able to estimate the amount of secondary color that might

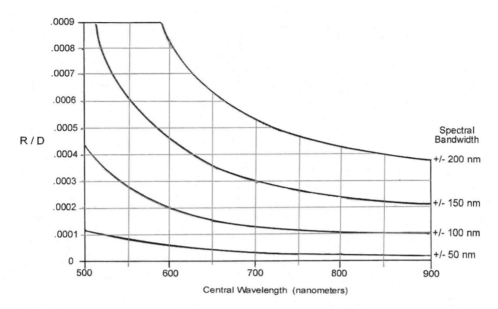

Figure 6.10 The blur-spot radius *R* due to residual secondary color in a conventional achromat will depend on the central wavelength, the spectral bandwidth, and the lens diameter *D*. Using the above data, it will be possible to estimate the blur-spot radius and the approximate maximum lens resolution 1/*R*.

remain after all primary aberrations have been corrected. Figure 6.10 shows a plot that makes it possible to estimate the residual blur-circle radius due to secondary color, assuming we know the diameter of the lens aperture, the central wavelength, and the spectral bandwidth that is being imaged.

When an optical system contains a series of lenses, all of which have been corrected for primary color, it is valid to add the lens diameters (based on the size of the axial light bundle) and use the sum of these diameters as the value *D* when using Fig. 6.10 to estimate residual secondary color. The conventional submarine periscope is an extreme example of this condition. In that case we might typically have an objective lens that is 42 mm in diameter, followed by six relay lenses, each of which is 100 mm in diameter. All of these lenses are achromatized, i.e., corrected for primary color over the visual spectral band. If we assume that spectral band to be 560, +/–50 nm, we will find from Fig. 6.10 a value for *R*/*D* that is 0.00008. Knowing the cumulative lens diameter to be 642 mm, we can then solve for the blur spot radius $R = 0.00008 \times 642 = 0.050$ mm. The resulting resolution limit at the final image will be about $1/0.050 = 20$ cycles/mm. This periscope design will contain an eyepiece with an EFL of 71.5 mm to view the final image with a magnification of $254/71.5 = 3.5\times$. This eyepiece magnification will boost the visual resolution capability to about 25 cycles/mm. From this we can conclude that the image quality of the periscope optics is consistent with the capability of the typical viewer's eye. It has been found that, when this optical system is to be used for a high-quality photographic application, the secondary color does become a limiting factor and a color filter will be required to reduce the spectral bandwidth in order to take advantage of the film's resolution capability.

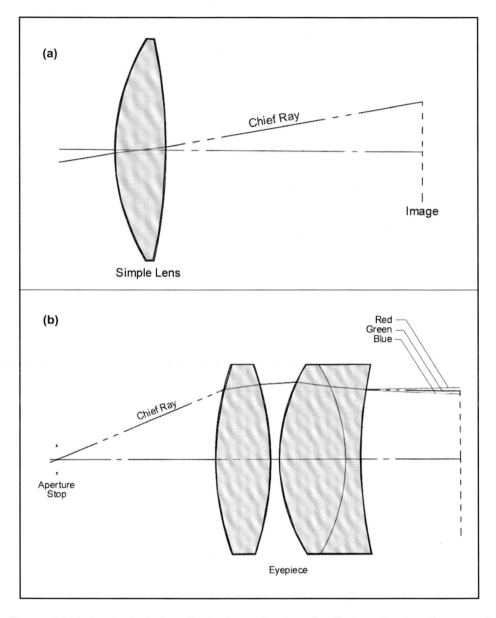

Figure 6.11 Lateral color is the off-axis chromatic aberration that results when the amount of chief-ray refraction varies as a function of wavelength. When little chief-ray refraction occurs, as with the simple lens (a), there will be little lateral color. When considerable chief-ray refraction occurs, as with an eyepiece (b), there will be a considerable amount of lateral color.

6.9 Lateral Color

The second chromatic aberration (and the last of the seven primary lens aberrations) is *lateral color*. For an on-axis light bundle, the central ray of that bundle will coincide with the optical axis of the lens. For off-axis light bundles the corresponding central ray is referred to as the chief ray. The height of the chief ray at the image plane defines image size. When lateral color is present in the lens system, the refraction of the chief ray will vary as a function of wavelength, causing each wavelength to be imaged at a slightly different height on the image plane. The result will be a chromatic radial blur for the off-axis image points. In the case of a simple lens with the chief ray passing through the center of the lens, there will be minimal refraction of that ray and therefore minimal lateral color. Any lens system that is nearly symmetrical about the point where the chief ray crosses the optical axis (the aperture stop) will have little or no lateral color because the tendency will be to have lateral color cancel as it passes through the front and rear halves of that lens.

On the other hand, the eyepiece is a classic example of a lens form that has large amounts of chief ray refraction that is not symmetrical about the aperture stop. As a result, in most eyepiece designs, lateral color will be found to be a major contributor to degradation of off-axis image quality. Figure 6.11 illustrates the chief-ray path, first through a simple lens and then through a typical eyepiece. The presence or lack of chief-ray refraction and the resulting lateral color are shown for each case.

6.10 Aberration Curves

Exposure to various optics textbooks and reference material dealing with optical systems and lens designs will generally lead to the presentation of aberration curves in one form or another. Aberration curves represent a method of plotting lens performance data that can be very useful. In this section we will review and discuss the set of typical aberration curves shown in Fig. 6.12, highlighting some of the more interesting and useful information that can be derived from them. These ray-trace analysis curves are representative of those generated by the OSLO software package. Other programs will generate different formats, but the basic information will be the same.

In the left-hand portion of Fig. 6.12, aberration curve data has been generated by ray tracing three tangential and three sagittal fans of rays through the lens to its image surface. In this case, each of the tangential ray fans contains nine rays that are equally spaced from the top to the bottom of the aperture. Each sagittal ray fan contains four rays, equally spaced from the edge of the aperture to the center. For the sagittal case, left to right symmetry makes it unnecessary to trace both sides of the aperture. Tangential and sagittal aberration curves are plotted for three field positions: on axis (AXIS), at 70% of the field (0.587 deg), and at 100% of the field (0.83 deg). (Note that in this case, the term "field" refers to the maximum field angle, which is half of the full field of the lens). From left

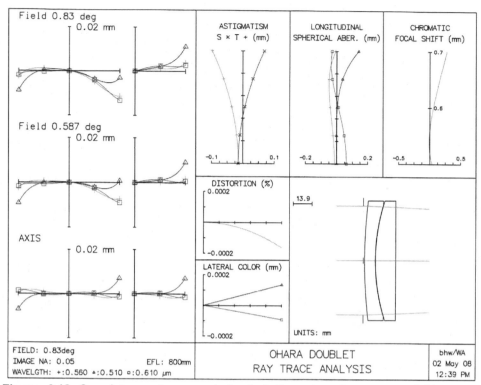

Figure 6.12 Complete ray-trace analysis, including aberration curves, field curves, distortion, lateral color, and a lens drawing. The scales for this output from OSLO can be edited by the user.

to right, each tangential plot represents rays from the bottom to the top of the aperture. The vertical dimension plotted is the intersection point for each ray at the image surface, representing residual errors in the vertical direction. Curves are plotted for the three principal wavelengths. For a perfect lens with no residual aberrations, each curve would be a horizontal line coincident with the reference axis. Similarly, for the sagittal plots, from left to right each point represents rays from the center to the edge of the aperture and the deviations plotted represent residual errors in the horizontal direction. The interpretation of aberration curves will generally be left to the experienced lens designer during the actual lens design process.

In the central part of Fig. 6.12 there are three sets of plotted data. At the top are the tangential and sagittal field curves. In the center is a plot of residual distortion. At the bottom residual lateral color plots will be seen. On the right side (top) of Fig. 6.12 there are curves showing longitudinal spherical aberration and chromatic focus shift. Finally, in the lower right corner of Fig. 6.12 there is a scale drawing of the lens under analysis.

Prior to the availability of the modern PC and software packages, the lens designer would go to great lengths to minimize the number of rays to be traced through a lens system during analysis. Useful results could often be obtained by tracing as few as seven total rays. Today, of course, that problem no longer exists. For example, with a single keystroke, in a fraction of a second, the aberration curves shown in Fig. 6.12 will be generated and plotted, tracing several hundred rays and generating an accurate lens picture. A quick review of Fig. 6.12 tells the experienced optical designer that spherical aberration is well corrected, as is primary color. Off-axis aberrations, including coma and astigmatism, will dominate off-axis image quality. Residual distortion and lateral color are negligible. Longitudinal spherical aberration is minimal and primary chromatic aberration has been corrected such that focus at 0.50 and 0.60 μm will coincide. The lens picture confirms a reasonable doublet design with good thickness-to-diameter ratios. It is a simple matter, and often quite useful, to edit the ray-trace analysis output format by altering the number of field positions and the plotting scale on any of the outputs.

6.11 Point-Spread Function Analysis

There are several other routines that can be used to evaluate the impact that residual aberrations will have on final image quality. One very valuable image-quality analysis tool is the point-spread function (PSF). In order to generate PSF data it is necessary to trace several thousand rays from one object point through the lens system and onto the image surface. All design wavelengths must be considered, and the resulting geometric spot must be combined with the diffraction spot to yield a very accurate representation of the actual performance of the lens. Applying the PSF tool in OSLO to the semiapochromatic doublet referenced in Chapter 5, Fig. 5.15, we will generate the PSF data shown in Fig. 6.13. From bottom to top on the left side of that figure there are three-dimensional plots of the resulting PSFs for the on-axis case, the 0.71 field, and the full-field points. Because this lens is nearly diffraction limited, the PSFs demonstrate the typical diffraction pattern, consisting of a very bright central spot surrounded by a series of rapidly diminishing concentric rings. In the three central boxes we have accurate cross sections through each PSF in both the X and Y direction. The maximum height of each of these plots is indicative of the maximum illumination level of each spot. This relative illumination value is referred to as the *Strehl ratio*. For lenses that are well corrected such as this, the Strehl ratio is a good measure of image quality. A Strehl ratio from 1.0 to 0.80 indicates a system with image quality that is essentially diffraction limited.

Finally, on the right side of Fig. 6.13, we have two plots indicating the ensquared (top) and encircled (bottom) fractional energy for the three field points. While encircled energy is the traditional presentation, in today's field of detectors with square and rectangular pixels, the ensquared energy numbers have become increasingly important.

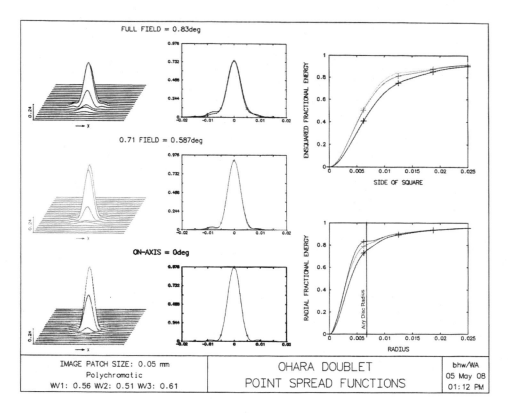

Figure 6.13 Point-spread functions for the semiapochromatic doublet that was described in Fig. 5.15.

The equation for the radius of the first dark ring in the Airy disk is

$$R = 1.22\,\lambda\,(f/\#).$$

For this doublet, the radius will be $R = 1.22 \times 0.00056 \times 10 = 0.0068$ mm. That Airy-disk radius has been indicated on the encircled energy plot. Note that it coincides with the point where the on-axis plot flattens out, indicating the first dark ring in the pattern and confirming diffraction-limited on-axis image quality.

6.12 Review and Summary

This completes our review of the seven primary lens aberrations, along with a brief look at how these aberrations can be identified and evaluated using ray-trace data (aberration curves) and PSF analysis. It is the function of the optical designer to evaluate the impact that residual aberrations will have on ultimate system performance and to adjust the configuration of the lens system such that satisfactory performance will result. The information presented in Fig. 6.14

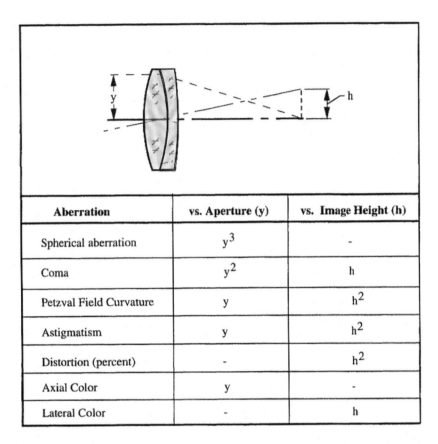

Aberration	vs. Aperture (y)	vs. Image Height (h)
Spherical aberration	y^3	-
Coma	y^2	h
Petzval Field Curvature	y	h^2
Astigmatism	y	h^2
Distortion (percent)	-	h^2
Axial Color	y	-
Lateral Color	-	h

Figure 6.14 The blur circle diameter (spot size) due to the primary lens aberrations will vary as a function of the aperture radius *y* and the image height *h*.

shows the relationship of the seven primary aberrations to the size of the lens aperture *y* and the field of view *h*. For example, when the lens aperture is doubled, the spherical aberration will increase by a factor of 8×, coma by 4×, and field curvature and axial color by 2×. Distortion and lateral color will not be affected.

Discussion of basic aberrations—understanding and evaluating them—has once again shown quite dramatically the impact that the modern PC and available optical design software has had on the scope of work and design responsibilities that are shared today by the optical engineer and the lens designer.

Chapter 7
Optical Components

7.1 Introduction

All optical instruments or systems contain a series of optical components. Each of these components has a specific function involving the collection, refraction, reflection, or deviation of the light coming from the object. This chapter is intended to give the reader a basic level of familiarity with this family of optical components. Armed with this information, the optical engineer will be better equipped to specify, design, and analyze a broad range of optical systems. The three optical components most commonly encountered are: the lens, the mirror, and the prism. Each of these will be described here in some detail, including a discussion of typical applications and performance limitations.

7.2 Lens

The simple lens is a primary component in nearly all optical systems. When a bundle of light rays emerges from a point on the object and is incident on a positive lens, those light rays will be refracted by the lens and they will be brought to a focus, forming an image of the original object point. A simple lens element can be described and thoroughly analyzed if the following detailed parameters are known:

- lens material
- radius of both lens surfaces
- lens center thickness
- lens diameter.

In those cases where a substantial spectral bandwidth is to be imaged, it will be found that, due to dispersion of the glass, the image quality of a single lens element will rarely be adequate. In most cases, as touched on in the discussion of axial color (Sec. 6.8), the single-lens design must be achromatized. This will result in a two-element, or doublet, form of the simple lens. The two elements of the doublet may be cemented, or they may be air spaced. It is the designer's task to evaluate the pros and cons of each approach and to select the doublet form that will work best in the particular application being considered. In the case of the cemented doublet, the manufacturing tolerances on the cemented surfaces will be less stringent and antireflection (AR) coatings will not be required. Also, if the doublet is cemented the need for a lens spacer will be eliminated. On the other hand, the air-spaced doublet will generally deliver superior image quality while eliminating the cementing operation. Depending on the lens diameter, lens cementing can be problematic, producing mechanical strain and deformation of the elements that can lead to seriously degraded image quality. While none of these factors is conclusive, they should all be considered when deciding between the cemented and the air-spaced doublet form.

One final configuration in the genre that can reasonably be referred to as the simple lens would be the doublet with a split-crown element. The achromatic doublet will be made up of a positive lens element made from a *crown* glass type and a negative element made from a *flint* glass. In general, crown glasses will be have a lower index of refraction and less dispersion than flint glass. When it is found that the image quality of an achromatic doublet is limited by residual spherical aberration, it is often possible to improve that image quality substantially by splitting the crown into two elements, each having about one-half of the lens power found in the original single-crown element.

7.2.1 Typical lens application

To illustrate these simple lens forms and to explore their limitations, we will consider a typical application and generate examples of how these simple lens types might perform. For example, assume we are to generate the design for a simple, low-power telescope to be used for visual observations, i.e., a 10× monocular.

Based on the desired final size and weight of the instrument, we might conclude that an objective lens with a diameter of 50 mm and a focal length of 400 mm would work well. Our first step in the design process would be to select and analyze a simple crown-glass element with these dimensions to determine its image quality.

Using a simple optical design software package (such as OSLO), we can start by looking at a 50-mm-diameter single lens element made from BK7 glass, with a focal length of 400 mm, and a 2-deg half-field angle. Figure 7.1 shows computer output for such a lens with minimized spherical aberration. From the MTF data shown we can conclude that, due to the uncorrected axial color, this lens will resolve a maximum of about 10 cycles/mm at its image surface. Because the telescope is to have a magnification of 10×, we can conclude that a 40-mm eyepiece will be used in conjunction with the 400-mm objective lens. This eyepiece will magnify the eyes' acuity, increasing the visual resolution limit at the objective image to about 50 cycles/mm. Knowing this, we conclude that the 10 cycle/mm resolution of the objective lens will be unacceptable.

Next, we convert the single-element objective lens to a cemented achromatic doublet form. Data for the resulting design is shown in Fig. 7.2. Here we have the

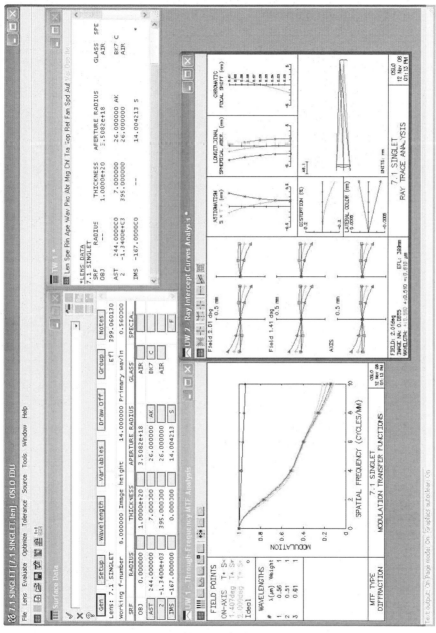

Figure 7.1 A simple lens with a 50-mm diameter and a 400-mm EFL is selected as an objective lens for a 10× telescope. The lens described here is afflicted with chromatic aberration, limiting resolution to about 10 cycles/mm. The above window from OSLO shows lens data, aberration curves, and MTF data.

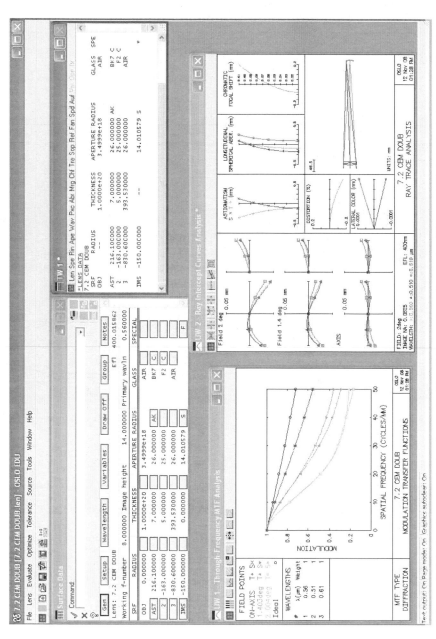

Figure 7.2 A simple cemented achromat with a 50-mm diameter and a 400-mm EFL is selected as an objective lens for a 10× telescope. The lens shown here is afflicted with residual spherical aberration, secondary axial color, astigmatism, and coma. All residual aberrations have been balanced to yield reasonable image quality. The above window from OSLO shows lens data, aberration curves, and MTF data.

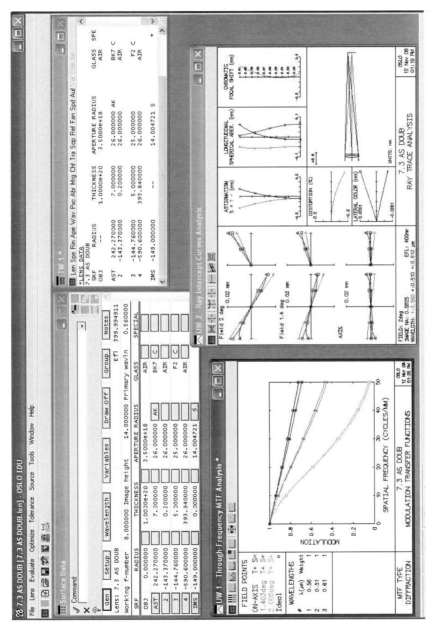

Figure 7.3 An air-spaced achromat with a 50-mm diameter and a 400-mm EFL is selected as an objective lens for a 10× telescope. The lens shown here is afflicted with residual secondary color and astigmatism. These residual aberrations have been balanced to yield reasonable image quality. The above window from OSLO shows lens data, aberration curves, and MTF data.

same criteria of 50-mm diameter, 400-mm focal length, and a half-field angle of 2 deg. In this design the residual spherical aberration and axial color are balanced to produce reasonable on-axis image quality. The MTF data indicates that at the limiting visual frequency of 50 cycles/mm, on-axis image contrast will be about 55%. For off-axis image points, residual coma and astigmatism will drop the contrast down to about 25%. This cemented doublet design would represent a fairly good solution, although the residual coma (apparent from the inverted "U" shape of the off-axis aberration curves), which subtends about a 3 arcmin angle to the eye, might be problematic. It will be worthwhile to examine the air-spaced doublet approach to see if a further improvement is possible.

Figure 7.3 shows the computer output for the air-spaced doublet design. From the MTF data shown we can see that the on-axis correction is much improved over that of the cemented design in Fig. 7.2. The MTF data indicates that the on-axis image contrast at 50 cycles/mm has been increased to about 70%, essentially equal to the maximum possible due to diffraction effects. The aberration curves show that residual coma has been eliminated from the air-spaced design, leaving astigmatism as the dominant off-axis aberration. Referring again to the MTF curves in Fig. 7.3, we see that the image contrast at 50 cycles/mm, for the 0.7 and 1.0 field points, is about 45% and 5% respectively. This represents a very good solution for the visual optical system being considered. Astigmatism is a less objectionable residual aberration, in that small amounts can be accommodated by the eye; this is not true of coma.

It was mentioned earlier that the next level of complexity for an objective lens might involve the splitting of the crown element into two elements, each with reduced lens power. The advantage of the implementation of this approach is a reduction in residual spherical aberration. In this case we have found that the on-axis image quality of the air-spaced doublet is essentially diffraction limited, making any substantial improvement literally impossible. We can conclude that consideration of the split-crown approach in this case would not be reasonable.

In summary, for this example of an objective lens for a low-power telescope, we have found that achromatization (color correction) is essential in order to have acceptable on-axis image quality. In addition, we have found that by introducing the variables of an air-spaced doublet design we are able to balance all residual aberrations resulting in a design with diffraction-limited, on-axis image quality, very good image quality over 70% of the field, and satisfactory image quality out to the edge of the full field. The design of the 10× telescope will be completed by the selection of a 40-mm eyepiece from an established catalog source.

7.2.2 Detail lens drawing

At the completion of the design phase, the computer software package will provide a complete set of numerical data describing in detail the finished lens form. This data will include the optical glass types being used, the radius on each lens surface, the axial thickness of each lens element, and all air spaces and lens

```
*PARAXIAL CONSTANTS
    Effective focal length:    400.00000    Lateral magnification:    -3.9999e-18
    Numerical aperture:          0.06250    Gaussian image height:       13.96813
    Working F-number:            8.00000    Petzval radius:            -559.34364


*LENS DATA
AIR SPACED DOUBLET
  SRF       RADIUS       THICKNESS    APERTURE RADIUS       GLASS  SPE  NOTE
  OBJ        --          1.0000e+20    3.4921e+18           AIR

  AST     242.27000 V     7.00000      26.00000 A           BK7  C
   2     -143.37000 V     0.20000 V    26.00000             AIR

   3     -144.76000 V     5.00000      25.00000             F2   C
   4     -590.60000 V   393.84000      26.00000             AIR

  IMS    -149.00000 V       --     V   14.00000                     *


*WAVELENGTHS
CURRENT  WV1/WW1     WV2/WW2     WV3/WW3
   1     0.56000     0.51000     0.61000
         1.00000     1.00000     1.00000


*REFRACTIVE INDICES
  SRF    GLASS         RN1         RN2         RN3        VNBR
   0     AIR         1.00000     1.00000     1.00000       --
   1     BK7         1.51803     1.52077     1.51591    106.59032
   2     AIR         1.00000     1.00000     1.00000       --
   3     F2          1.62262     1.62851     1.61821     60.43499
   4     AIR         1.00000     1.00000     1.00000       --
   5     IMAGE SURFACE
```

Figure 7.4 Computer output from OSLO, describing the nominal characteristics of the 400-mm EFL, 50-mm diameter air-spaced doublet objective lens.

surface clear apertures. It is an important function of the optical engineer to convert this raw numerical data into detail drawings that will permit the optical shop to manufacture each lens. The detail drawing must include all critical basic data, along with all related manufacturing tolerances. Certain physical characteristics, such as lens diameter, will be determined by the mounting methods to be employed along with other mechanical design considerations. In this section we will describe the process of converting the lens design data for the 400-mm air-spaced doublet to the detail drawing format.

Figure 7.4 shows the computer data printout for the air-spaced doublet referenced in Fig. 7.3. While all nominal data describing the lens are present, the format must be modified and tolerances must be added in order to produce a document that will be usable by the optical manufacturing operation. Figure 7.5 is a finished working drawing of the same air-spaced doublet. While numerous optical drawing formats are used throughout the industry, the generic format shown here contains all of the information needed for the manufacture of a satisfactory pair of lenses. Much work has been done in recent years directed toward the generation of a universal drawing standard. Procedures in place at most design and build facilities today will have computer-aided design personnel who will take the output from the optical engineer (in a form similar to Fig. 7.5) and convert this drawing to the officially acceptable, final format.

A key component of any lens drawing will be a picture of the lens, preferably drawn to scale, which will convey an accurate representation of the lens shape and proportions. Next, all basic lens dimensions such as diameter and thickness are added. Each optical surface is given an identification number (R_1, R_2 ....) and all relevant optical information for these surfaces is presented in tabular format.

The subject of tolerance analysis is critical to the manufacturing phase. The approach used to determine tolerances will vary, depending on the complexity of the design and the analytical tools that are available. In this particular example, it is a fairly simple matter to use software to introduce small changes to the various lens parameters, determine the effect of each change on image quality, and then settle on an acceptable value for each manufacturing tolerance. Tolerances on radius, thickness, and element wedge were established in this case. For each optical surface, tolerances on radius, sphericity, and cosmetic finish (scratch/dig) are tabulated. The final required image quality—diffraction limited, on axis—will dictate the surface power and irregularity (4 × 0.5 fringe). The final application, in this case an objective lens for use in a visual instrument, will dictate the required surface cosmetic quality (80-50 scratch/dig). In special cases, such as a high-power laser system, the cosmetic quality of the polished surfaces will become much more critical. The military standard on optical components (MIL-O-13830) should be reviewed and carefully considered by all involved in the area of optical tolerances and quality.

The drawing in Fig. 7.5 also contains a separate table that specifies the types of optical glass to be used for each element. A set of general notes is added to cover other information such as AR coatings and cementing, if required. In the surface data table the free-aperture diameter for each surface is given. This is important because coatings and optical quality notes will not apply outside of that diameter.

I consider the most important rule that should be applied to the preparation of any optical component drawing is that it should be easily understood and completely free of ambiguity. If any of the items covered on the drawing is open to interpretation, the drawing has not done its job and the resulting manufactured part will most likely not perform up to expectations. One final point: each drawing must be assigned a unique drawing number that will allow it to be easily and accurately identified without confusion as work on the part progresses.

7.3 Mirror

Any optical component whose primary function is to reflect incident light is called a *mirror*. In the design of a mirror, two factors are of special importance: the shape of the reflecting surface and the reflectivity characteristics of that surface. The simplest and most common mirror is the plano, or flat mirror. The flat mirror serves to deflect the optical path while having no effect on the

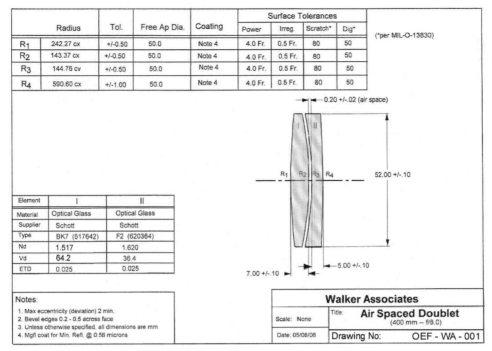

The table portion:

	Radius	Tol.	Free Ap Dia.	Coating	Surface Tolerances				
					Power	Irreg.	Scratch*	Dig*	(*per MIL-O-13830)
R_1	242.27 cx	+/-0.50	50.0	Note 4	4.0 Fr.	0.5 Fr.	80	50	
R_2	143.37 cx	+/-0.50	50.0	Note 4	4.0 Fr.	0.5 Fr.	80	50	
R_3	144.76 cv	+/-0.50	50.0	Note 4	4.0 Fr.	0.5 Fr.	80	50	
R_4	590.60 cx	+/-1.00	50.0	Note 4	4.0 Fr.	0.5 Fr.	80	50	

Element	I	II
Material	Optical Glass	Optical Glass
Supplier	Schott	Schott
Type	BK7 (517642)	F2 (620364)
Nd	1.517	1.620
Vd	64.2	36.4
ETD	0.025	0.025

0.20 +/-.02 (air space)

52.00 +/-.10

5.00 +/-.10

7.00 +/-.10

Notes:
1. Max eccentricity (deviation) 2 min.
2. Bevel edges 0.2 - 0.5 across face
3. Unless otherwise specified, all dimensions are mm
4. Mgfl coat for Min. Refl. @ 0.56 microns

Walker Associates

Title: **Air Spaced Doublet**
(400 mm -- f/8.0)

Scale: None

Date: 05/08/08

Drawing No: OEF - WA - 001

Figure 7.5 Complete detail drawing describing both elements required to make up the 400-mm EFL, 50-mm diameter air-spaced achromatic doublet.

reflected light in terms of convergence, divergence, or optical aberrations. Because of this characteristic, the optical system can be designed ignoring any flat mirrors that may be introduced into the system at some later stage. In nearly all optical design applications the reflecting surface will be the first surface of the mirror. As a result, the choice of material for the mirror substrate will not be critical.

Another factor that cannot be ignored is the effect that each mirror will have on the orientation of the image produced by the optical system. While a precise and rigorous evaluation must be made in all cases, a general rule of thumb for system design is that an odd number of flat mirrors within the system will result in a reversed image . . . that will not be acceptable. Conversely, an even number of mirrors will generally result in a correct image orientation that will be acceptable.

To demonstrate this phenomenon, let's consider the 10× visual telescope that was introduced during the discussion of lenses. Assume for the moment that we want the telescope to contain a 90-deg bend, so that in order to look straight ahead, we would need to look straight down into the telescope eyepiece. It can be seen from Fig. 7.6 that the addition of a flat mirror at a 45-deg angle between the objective lens and its image will produce the desired 90-deg bend. However, we can see that the image being viewed will be reversed left-to-right, which would not be acceptable in most cases. For example, if the telescope were intended for

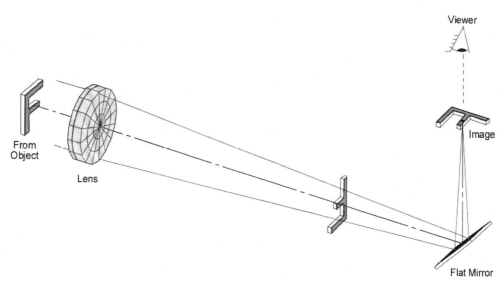

Figure 7.6 Placing a single flat mirror between the objective lens and its image reflects the light path upward. It also produces a left-to-right reversion of the image as it appears to the viewer. In most applications this is an unacceptable condition.

astronomical viewing of stars and planets, this image reversal would not be a serious problem. However, in the more likely case where the telescope was to be used for terrestrial viewing of birds, animals, or sporting events, the reversed image would be totally unacceptable. The most common solution to this image reversal problem is to use a roof mirror in place of the single flat mirror (see Fig. 7.7). Critical to the satisfactory performance of a roof mirror is the fact that the angle between the two mirrors that make up the roof must be held precisely at 90 deg. Any error in this 90-deg roof angle will result in the formation of a double image, which will adversely affect the final image quality of the instrument. Typically, a tolerance of from 1 to 5 arcsec will be required on the 90-deg angle of the roof mirror assembly. Because of this tight tolerance, the roof mirror is most often replaced by a roof prism, where the 90-deg roof angle can be more easily accomplished and will not be subject to assembly and alignment problems (more on roof prisms later in this chapter).

The specification of optical quality on a mirror will be related primarily to the shape of the surface and the type of reflecting material required. The mirror shown in Fig. 7.6 is typical of folding mirrors that might be found in any number of optical instruments. In this case, the surface must be flat. Since, as we stated earlier, the objective lens performance is to be diffraction limited, the mirror should not introduce an error to the converging wavefront that is greater than one-eighth wave. In order to determine the actual aperture size requirement for this mirror we can trace ray bundles for the on-axis case and for the upper and lower field points. In this case the field of view is 4 deg and the mirror is located at a point on the optical axis that is 300 mm from the last surface of the objective

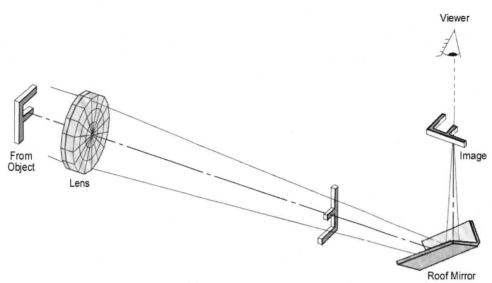

Figure 7.7 Placing a roof mirror between the objective lens and its image reflects the light path upward. It also eliminates the left-to-right reversion of the image that was produced by the single flat mirror shown in Fig. 7.6.

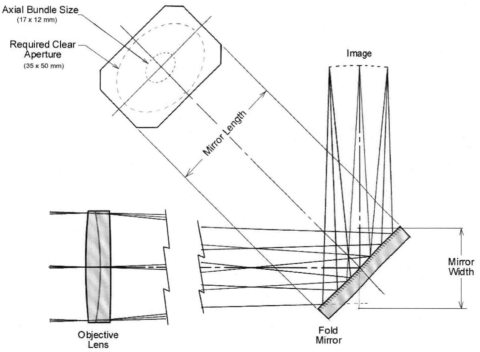

Figure 7.8 The required aperture size of a flat folding mirror located between an objective lens and its image will depend on the axial location of the mirror and the ray bundles from lens to image.

lens. With this information input to the lens analysis program, the ray trace shown in Fig. 7.8 can be generated. From this ray trace we can conclude that light from each object point will form an elliptical spot on the mirror surface that is about 17 × 12 mm, and the overall clear aperture requirement will be a larger ellipse that is 50 × 35 mm in size.

Figure 7.9 shows a detail drawing that will allow manufacture of the folding mirror for use in our telescope. The tolerance assigned to reflecting surface flatness is 0.25λ, where λ is 0.633 μm, a convenient wavelength for interferometric testing. Let's now consider how this tolerance will affect the spherical shape of the on-axis converging light bundle. We have found by ray tracing that the size of that bundle as it reflects from the mirror will be a 17 × 12-mm ellipse. Assume the likely case where the finished mirror surface has a cylindrical shape, perfectly flat in the cross section seen in Fig. 7.8, and concave by 0.25λ (0.00016 mm) in the direction in and out of the plane of the page. The mirror aperture and the axial bundle in this direction are 35 mm and 12 mm, respectively. If the mirror surface flatness error is 0.25λ over 35 mm, it will be equal to $(12/35)^2 \times 0.25\lambda = 0.029\lambda$ over the 12-mm bundle dimension. Because this is a reflecting surface at a 45-deg angle, the resulting reflected wavefront error will be 1.4× the surface error, or 1.4 × 0.029λ = 0.041λ. Any wavefront error less than one-eighth (0.125) wave will be consistent with our goal of diffraction-limited performance for the telescope; thus, this tolerance will be acceptable.

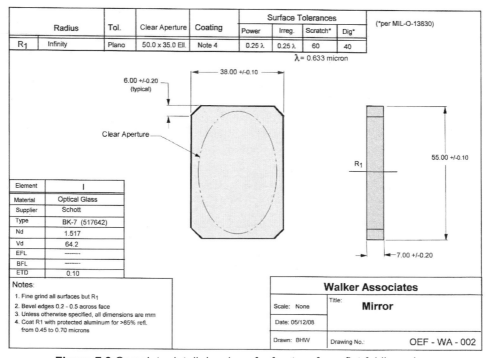

Figure 7.9 Complete detail drawing of a front-surface, flat-folding mirror.

In choosing a reflecting material for the mirror surface, the major considerations will be percent reflectivity, the wavelengths to be reflected, material durability, and cost. In this case where we are dealing with a conventional visual instrument, a polished glass substrate with a deposited coating of protected aluminum will generally be satisfactory. Such a coating will reflect about 85% of light within the visible spectrum, and it will be quite durable and cost effective. Note that for precision optical applications, the reflecting surface must be the first surface of the mirror. Although making the second (back) surface the reflector does offer the advantage of protecting the mirror surface from damage, the reflection from the uncoated first surface will cause a ghost image. Also, any imperfections in the glass substrate itself will degrade performance.

Figure 7.9 shows a typical detail drawing of the mirror designed for this application. Making the overall mirror shape rectangular rather than elliptical results in a component that is more easily produced, while at the same time, this shape provides space for mounting hardware outside of the mirror's clear aperture.

On the subject of mounting the mirror, it is a common error to produce a mirror of high quality (flatness) and then to destroy the performance of that mirror by mounting it such that its surface is mechanically distorted. Extreme care must be taken in the mounting of all optical components, especially mirrors and prisms, to ensure that no mechanical stresses are introduced that will disturb the quality (flatness) of the reflecting surface. The potential for a problem is best illustrated by considering the fact that, while a very tight mechanical tolerance might be in the region of +/–0.0001 in., for our mirror it is essential that the reflecting surface remain flat, after mounting, to within +/–0.0000055 in., a ratio of nearly 20:1.

7.3.1 Common reflecting coatings

The basic aluminum coating with a protective layer is the most common and cost-effective choice for a reflecting surface in most optical applications. All mirror coatings will be vacuum deposited, with layer thicknesses that range from one micron to a fraction of a wavelength in thickness. In the visible portion of the spectrum, from 0.45 to 0.70 µm, an aluminum-coated mirror will have an average reflectance of about 88%. It is possible to improve this reflectance to about 94% by adding several enhancement layers of dielectric material on top of the aluminum.

As is shown in Fig. 7.10, aluminum coatings have an inherent dip in their reflectance, to about 80%, at a wavelength of 0.80 µm. For visual systems this is not a problem. However, many modern electronic imaging systems will have detectors with peak sensitivity around this wavelength. If several reflecting surfaces are to be used in such a system, it will be advisable to change the design to call for a silver coating, rather than an aluminum reflecting surface. While

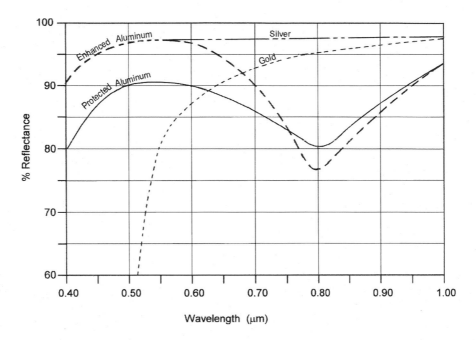

Figure 7.10 A number of materials are used for the reflective coatings on optical mirrors. Shown here are the reflectance curves for the four most common coatings.

silver offers a clear advantage in terms of reflectivity, it does tend to be more expensive and also more susceptible to damage due to environmental conditions. This durability factor must be considered by those responsible for system design and preparation of coating specifications. Last among the more common reflective coating materials to be discussed here is gold. Gold coatings are frequently used for mirrors in infrared systems. The average reflectance of gold over the infrared band extending from 0.80 μm (near IR) to 12.0 μm (far IR), will be >96%. As its yellowish appearance indicates, a gold coating does not reflect uniformly across the visible spectrum, falling to essentially zero at a wavelength of 0.50 μm. Reflectance curves for aluminum, silver, and gold mirror coatings are shown in Fig. 7.10.

7.3.2 Other mirror configurations

The flat mirror shown in Fig. 7.9 is the most common mirror form found in optical systems. In this case the only function of the glass substrate is to provide a stable, polished platform for the reflective coating of vacuum-deposited aluminum. As I said earlier, because the light reflects directly from the aluminum, never entering the glass, this is referred to as a *front surface* mirror. This front surface configuration is also often found in mirrors with concave and convex configurations. These curved mirror surfaces may be spherical, but are often aspheric to facilitate the correction of aberrations.

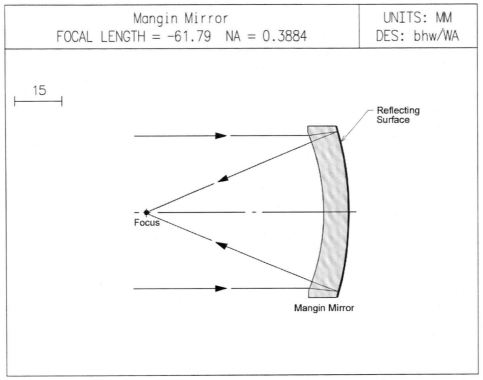

Figure 7.11 A Mangin mirror combines a negative meniscus lens and a concave mirror into a common element for purposes of aberration correction.

An exception to the front surface condition will be found in the *Mangin mirror*. As shown in Fig. 7.11, the Mangin mirror is a combination of a weak meniscus lens and a concave mirror, both combined into a single optical element for purposes of aberration correction. The Mangin mirror is occasionally used as the primary reflector in the design of an astronomical telescope. The substrate material in a Mangin mirror must be of the same optical quality that would be found in a lens used in a similar application.

The polished glass substrate with a vacuum-deposited reflective coating used to be the configuration of choice for use in most optical systems. Recently, in the 1970s two new processes were developed that offer alternative approaches to the optical designer: replicated and micromachined mirrors. In both cases the major advantage offered is that the mechanical means for mounting the mirror can be made an integral part of the mirror itself.

In the case of the replicated mirror, there is also a substantial cost advantage when large quantities are involved. For the production of a flat, front-surface mirror, the replication process begins with a master flat that has been polished to at least twice the precision required of the final mirror. As shown in Fig. 7.12, the first step in the replication process is to cover that master with a parting layer of nonstick material, such as silicone, that will not adhere to adjacent layers. This is

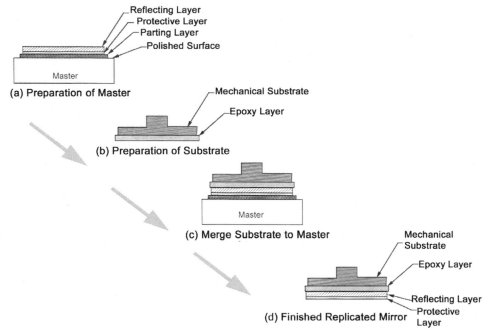

Figure 7.12 The procedure illustrated here shows how a replicated mirror is produced based on the use of a master surface. A major benefit will be the inclusion of the mechanical substrate into the finished mirror.

followed by the vacuum deposition of the required reflective coating, in reverse order, onto the parting layer. Meanwhile, the mechanical mirror substrate is fully machined, including the flat surface, to normal mechanical tolerances, and then the part is fully stress relieved. A thin layer of epoxy is then applied to the flat surface of the mechanical mirror substrate. The epoxy layer is then brought into intimate contact with the prepared master. After the epoxy has cured, the finished part may be separated from the master. The reflective coating now becomes an integral part of the mechanical mirror substrate. The replication process derives its name from the fact that the final reflecting surface will be a precise replica of the polished surface on the master. Depending on sizes involved, additional savings can be realized by producing several replicated mirrors simultaneously, using a common oversized master.

Micromachining, or diamond-point machining, is a very precise method of machining metal to tolerances that are consistent with the requirements of optical components to be used in infrared systems. This limitation to IR applications arises because the surface accuracy required will be a function of the wavelength that is involved. In most IR systems that wavelength will be from 10 to 20 times larger than the wavelength of visible systems. While micromachined surfaces must be produced individually, much like traditional optics, it is possible to produce these surfaces using computer-controlled machines that are essentially automatic and very precise. Another major advantage of the process is that these machines can be programmed to produce aspheric surfaces at little or no additional cost relative to spherical surfaces. The introduction of aspheric surfaces into a design will make possible a better performing system, often with fewer optical elements, resulting in substantial cost savings. In addition to metal mirrors, there are several IR refracting materials that can be micromachined. This makes it possible to produce aspheric IR lenses with all of the advantages that come with the micromachining process.

The micromachining process has been used to produce mirrors for visual optical systems. Unfortunately, the very fine residual pattern left on the surface by the micromachining process has a tendency to cause problems with scattered light and diffraction effects in systems operating at visible wavelengths. While it is possible to eliminate the residual machining marks by postpolishing the micromachined surface, this approach eliminates most of the advantages of the micromachining process.

7.4 Prisms

Like the mirror, the *prism* is an optical component most often used to deviate or displace the optical axis. Probably the most common prism form is the *right-angle prism* that would typically be used to replace a flat mirror such as the one shown in Fig. 7.6. There are several advantages to substituting a prism for a mirror in this application. First, the prism will be easier to mount because it is a much more rigid component than the mirror. Secondly, in the prism the reflective coating will be protected from physical or environmental damage. Once again, these advantages do not come without trade-offs. The prism must be made from optical glass that is of high optical quality, while the substrate quality of the front surface mirror is relatively unimportant. To produce a quality right-angle prism, three surfaces must be polished flat to a high quality optical finish, while for the front-surface mirror only one quality surface is needed. Finally, the prism will have an effect on image location and quality when it is located in a converging light bundle, as is the case in this example.

We will examine in some detail the effect that the addition of a right-angle prism will have on our previously discussed achromatic doublet (Fig. 7.5). This objective lens has a focal length of 400 mm and an aperture diameter of 50 mm. Analysis of this lens shows that the on-axis image quality is limited by about one-quarter wave of residual focus and color errors. We will now add a folding right angle prism and determine what must be done to maintain that level of image quality. Initial analysis shows that the prism has added about one-eighth wave of error. Using OSLO, the lens radiuses were modified (between 1 and 1.5%) and the residual errors were brought back to their initial values.

Also, inserting the prism will cause the axial location of the image surface to shift away from the lens by the following amount:

$$\text{Focus shift} = (n - 1.0)(t/n),$$

where n is the prism index of refraction and t is the prism thickness.

In this case the index of refraction is 1.517 and the prism thickness is 40 mm. The resulting focus shift will be 13.63 mm. A useful rule of thumb, applicable to prisms (or windows) with an index of 1.50, is that the image will be shifted by about one-third the glass thickness. A typical right-angle prism is shown in Fig. 7.13.

As was the case with the flat mirror (Fig. 7.6), the right-angle prism, with its single reflecting surface, will cause a reversal of the image that may be unacceptable. In that case it would be appropriate to use the equivalent of the roof mirror (Fig. 7.7), which is a right-angle prism with a roof surface in place of the single flat reflecting surface. It will be recalled that a roof surface consists of a pair of flat reflecting surfaces that are set at a precise 90-deg angle to each other. A roof prism of this type is traditionally referred to as an *Amici prism*.

In the case of the right angle prism, the total path of the optical axis through the prism is about equal to the prism aperture diameter. In the case of the Amici prism, due to the roof surface, the total path through the prism will be about 1.7× the prism aperture diameter. In the example of our telescope objective, which requires an aperture of 40 mm, the total path through the prism would be about 68 mm and the focus shift would be one-third of that, or about 23 mm.

Figure 7.13 illustrates several other common prism types. The *Porro prism* is most commonly found in traditional prism binoculars. By introducing a pair of Porro prisms into the optical path behind the objective lens the image is brought to the correct orientation for straight ahead viewing through the eyepiece. The *dove prism* is often introduced as an image derotation device within an optical instrument. Many instruments are designed to have a fixed eyepiece and viewing point, while a scanning mechanism directs the line of sight to various azimuth bearings (see Fig. 7.14). As that figure shows, as the instrument scans 360 deg in azimuth, the image being viewed will appear to rotate through that same angle about the optical axis. A dove prism can be linked to the scan mechanism in such a way as to produce a counter-rotation, resulting in a stable, upright image regardless of the azimuth direction being viewed. The *delta prism* is a very compact, folded version of the dove prism. In order to achieve the required internal reflection, the delta prism must be made from a high-index glass (*n* >1.70). In the cases of the dove and delta prisms, note that the optical axis strikes the first prism surface at a large angle of incidence. Because of this, it is not possible to use either of these prisms in a converging light bundle. The incident light bundle must be collimated, i.e., all light rays in the axial bundle must be parallel to the optical axis. On the positive side, any prism that is placed in a collimated light bundle will not introduce aberrations into the optical system.

A third form of derotation prism is the *Pechan prism*. The Pechan is made up of two separate prisms with a small airspace between them. In this case the optical axis is perpendicular to the entrance face of the prism. As a result, the Pechan prism can be used in a converging light bundle, but its impact on residual aberrations must be taken into account. The final prism form shown in Fig. 7.13 is the *penta prism*. Its name is derived from the fact that the prism has five sides.

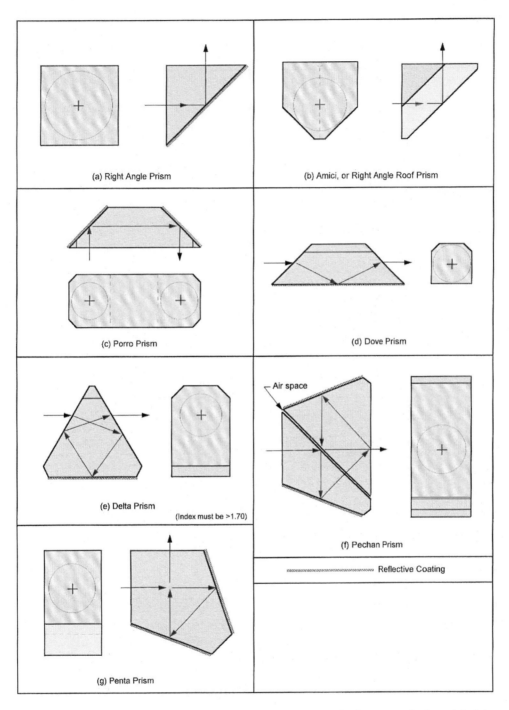

Figure 7.13 Various prism types are commonly used in optical system design: (a) right-angle prism, (b) Amici prism, (c) Porro prism, (d) dove prism, (e) delta prism, (f) Pechan prism, and (g) penta prism.

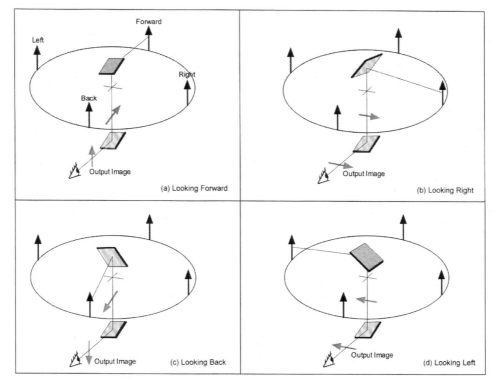

Figure 7.14 Illustration of the rotation of the output image created by an azimuth scan using the top mirror of a simple two-mirror periscope viewer.

The penta prism functions to deviate the optical axis through an angle of exactly 90 deg. Because of its unique construction, the penta prism is insensitive to small angles of rotation about the two axes perpendicular to the optical axis. The penta prism is often found in instruments such as the optical range finder, where a precise 90-deg angular deviation is critical. As was the case for the Amici prism, a reflecting surface in any of these prisms may be replaced with a roof in order to achieve the required image orientation.

7.4.1 Total internal reflection

Several of the prisms discussed in this section will function without the need for a reflective coating. The reflection that occurs within these prisms is referred to as *total internal reflection* (TIR). It will be helpful to review this topic and to establish a basic understanding of how it works. In Fig. 7.15 we see a block of glass with two polished faces labeled A and B. in the first case these faces are parallel to each other and perpendicular to the incident light ray. This is nothing more than a common optical window, and the light ray will pass through this window undeviated. Next, it can be seen that when the surface B is tipped relative to A, the part becomes a deviating prism, or a wedge, and the light ray is refracted at surface B in accord with the law of refraction. When the condition is

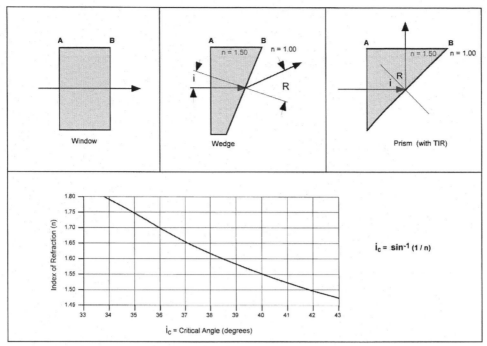

Figure 7.15 Total internal reflection (TIR) occurs when the sine of the angle of incidence within the glass is greater than 1/*n*. The point where this occurs is referred to as the critical angle.

reached where the sine of the angle of refraction *R* is greater than or equal to 1.0, then there will be no refraction, and all of the light will be reflected internally (TIR). The corresponding angle of incidence *i* is designated the *critical angle.*

TIR is a very basic and 100% efficient phenomenon. When it works, it works very well. However, there are several potential problems related to the application of TIR that must be carefully considered by the optical designer. First, the TIR surface must be kept totally free of contaminants such as moisture or fingerprints, which would defeat the TIR function. For this reason, the TIR surface cannot be used as the mounting surface for the prism. Also, the design must be carefully analyzed to ensure that none of the system's light rays are incident at less than the critical angle. Quite often the case will occur where the optical axis, along with some other light rays, meet the TIR criteria, but other light rays within the system do not. Because of these potential problems, it is often a reasonable design approach to apply a reflective coating to the surface, accept the 5 to 10% energy loss, and thus eliminate any risk. The graph shown in Fig. 7.15 contains a plot of the critical angle as a function of the index of refraction.

7.4.2 Typical prism application

Whenever a prism is added to an optical system, it is crucial that the designer very carefully analyze all of the factors that may be involved. Arguably the best reference on this subject is the prism section (13.7) of *MIL Handbook 141.*[1] A careful review and thorough understanding of that section, combined with the intelligent application of today's modern computer design tools, will allow the optical designer successfully to incorporate the required prisms into the optical system.

For example, assume that we wish to add a derotation prism in the back focal region of the 400-mm EFL objective lens shown in Fig. 7.5. Also, it may be important that the final design be as compact as possible. The logical selection to meet all of the design criteria will be a Pechan derotation prism [Fig. 7.13(f)]. The data sheet for the Pechan prism found in *MIL Handbook 141* tells us that the Pechan prism can be made from BK7, an ideal material for a relatively large prism. The same data sheet tells us that the optical path, along the optical axis, through the prism will be equal to 4.62× the prism aperture size, and that the mechanical length, from input face to output face, will be equal to 1.21× the prism aperture size. Based on the light bundle size in the back focal region of the doublet lens, a prism aperture size of 50 mm would be selected. Using the constants from the handbook we conclude that the total path through the prism will be 4.62 × 50 = 231 mm, while the physical length of the prism (from input face to output face) will be 1.21 × 50 = 60.5 mm. The first step in the design process is to insert a 231-mm-thick block of BK7 into the back focal region of our doublet. A quick analysis of image quality indicates that the on-axis blur circle, due primarily to axial color, will be increased from 0.005 mm in the original design to 0.015 mm with the glass block. This spot size is more than three times the diffraction limit, our goal for the final performance of this lens. As a result, it will be necessary to reoptimize the objective lens design to compensate for the aberrations introduced by the Pechan prism. Using the same optimization routines that were used to generate the original air-spaced doublet design, the lens shapes were modified to generate image quality that was equal to the starting design.

A positive feature of the Pechan prism is that, because the optical path is folded, the physical length of the prism is substantially less than the actual optical path through the prism. In this case, we have said that the physical length will be 60.5 mm, while the optical path through the prism will be 231 mm. The focus shift of the image plane due to the presence of 231 mm of BK7 will be 78.8 mm. When we combine all of the effects of the Pechan prism, we find that the overall length of the system will be reduced by 91.5 mm (see Fig. 7.16).

7.5 Review and Summary

This chapter has been intended to give the reader a feel for some of the basic optical components that will be found in today's optical systems. These component types have been identified as lenses, mirrors, and prisms. The simple positive lens may be the most frequently encountered optical component. Its various forms have been discussed, and the relative performance of the single

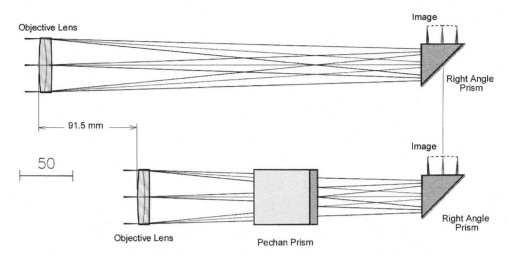

Figure 7.16 Illustration showing the effect on overall length due to adding a Pechan derotation prism to the back focus of a 400-mm EFL objective lens.

element versus cemented and air-spaced doublets have been evaluated. A hypothetical lens application has been presented for purposes of demonstrating design procedures and the evaluation of relative image quality. The preparation of detail drawings of optical components for use in the optical shop for manufacture has been covered in some detail.

The section on design and manufacture of mirrors for use in optical systems has included the most common mirror type—the flat, front-surface mirror—along with several more complex mirror forms. A brief discussion of replicated and micromachined mirrors has been included. Finally, the topic of prisms has been presented, including illustrations of the more common prism types, along with a discussion of their typical functions. A detailed example of the impact that the addition of a prism to an optical system might have on overall dimensions and image quality has also been presented.

All of this has been intended to help the reader to become more familiar and more comfortable with these basic optical components. Hopefully, this will prepare the optical designer to work with more advanced and comprehensive reference material, and, with today's optical design and analysis computer software, further develop essential design skills in these areas.

References

1. *Military Standardization Handbook: Optical Design* (MIL-HDBK-141), Defense Supply Agency, Washington D.C. (October 1965), as appears at: http://www.optics.arizona.edu/ot/opti502/MIL_HDBK_141.html.

Chapter 8
Basic Optical Instruments

8.1 Introduction

An *optical instrument* can be defined as a device used for observing, measuring, and recording optical data or information. It is often the responsibility of the optical engineer to design the optical system to be incorporated into such a device. This chapter will describe several basic instruments and discuss their applications and some of the special aspects of these instruments as they affect their design and manufacture. The most common optical instruments are those intended to enhance visual capability. This chapter will deal primarily with instruments of this type.

8.2 Magnifier, or Loupe

A magnifying glass (or loupe, from Old French, meaning an imperfect gem), is the simplest of optical instruments intended for the enhancement of visual capability. It is a device frequently associated with jewelers, usually taking the form of a simple positive lens. In use, the magnifier is held close to the eye, while the object to be viewed is brought to the focal point of the lens.

The average young and healthy human eye is capable of focusing from infinity, down to a minimum distance of about 250 mm (10 in.). This same average eye is capable of resolving a repeating high-contrast target with equal width black and white lines when each line subtends an angle of 1 arcmin or more. Most often, when viewing an object it is our intent to distinguish as much detail on that object as is possible. To that end we first bring the object as close as possible to the eye. When that closest distance is 250 mm, the smallest resolved element on the object—a detail that subtends an angle of one minute ($\tan\theta = 0.0003$)—will have an actual size of 250 mm × 0.0003 mm = 0.075 mm. If this resolved element is a part of a repeating pattern of equal-thickness parallel black and white lines, then each cycle (one black plus one white line) will have a thickness of 0.150 mm. The frequency of this finest resolvable pattern will then be 1/0.150 mm = 6.67 cycles/mm.

A magnifier is any positive lens with a focal length of less than 250 mm. The approximate magnification *M* provided by the lens is calculated by dividing its focal length into 250. For example, a 50-mm lens will provide a magnification of $M = 250/50 = 5\times$. This formula applies to the case where the object is placed at the focal plane of the magnifier lens and the virtual image being viewed appears at infinity. This condition allows comfortable viewing with the relaxed eye. If the object is moved just a bit closer to the lens so that the image is formed at the closest possible viewing point, a distance of 250 mm from the eye, then the magnification factor will be increased by an additional $1\times$.

These magnification formulas relate to a thin lens with that lens placed very close to the eye, both of which are hypothetical assumptions. Due to the ready availability of optical design software, it has become a relatively simple matter to analyze a real lens and to generate its actual magnification numbers. Let's assume a simple cemented doublet with a focal length of 50 mm and a diameter of 24 mm. We will further assume that we are using this lens to examine a coin that is 26 mm in diameter, as is shown in Fig. 8.1. When set for an infinite image distance, ray-trace analysis indicates that the tangent of the half angle to the edge of the image will be 0.265. Without the lens, if we look at the coin held at a distance of 250 mm, it will subtend a half angle whose tangent is $13/250 = 0.052$. Comparing the two half angles will give us the effective magnification of $0.265/0.052 = 5.1\times$. Next, if we reduce the distance from the lens to the object by about 9.4 mm, we will find that the image is formed at a point 250 mm from the eye. At that distance, the half angle to the edge of the image will now become 0.281. For this near point example the magnification will become $0.281/0.052 = 5.4\times$. These results represent real and accurate magnification numbers for this specific lens, including any residual distortion that may be present.

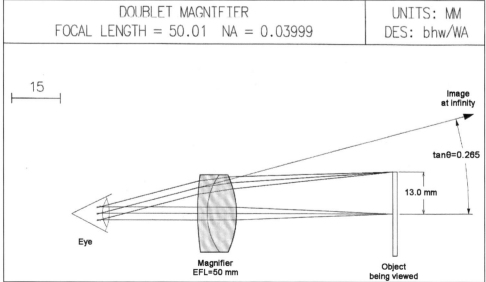

Figure 8.1 A 50-mm doublet used as a loupe, or magnifying glass. The image at infinity will subtend an angle about 5× larger than the angle subtended by the object when it is viewed with the naked eye at the near point of vision (250 mm).

PARAXIAL CONSTANTS

Effective focal length:	50.01261	Lateral magnification:	0.18638
Numerical aperture:	0.04292	Gaussian image height:	14.50000
Working F-number:	11.64921	Petzval radius:	-69.47448

LENS DATA

DOUBLET MAGNIFIER (NEAR)

SRF	RADIUS		THICKNESS	APERTURE RADIUS		GLASS
OBJ	140.00000		-250.00000	77.79750		AIR
AST	--		25.00000	2.00000	AS	AIR
2	87.20000	V	3.00000	12.00000		F2
3	21.17000	V	9.00000	12.00000		BK7
4	-28.44000	V	39.30000	12.00000		AIR
IMS	--	V	--	V	13.00000	

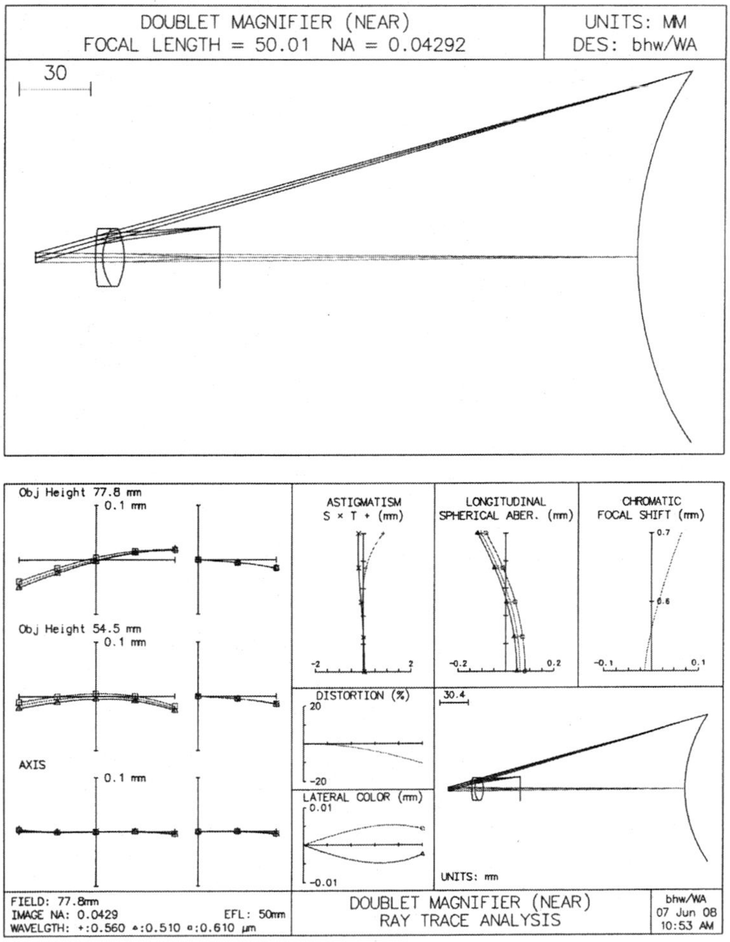

DOUBLET MAGNIFIER (NEAR)
FOCAL LENGTH = 50.01   NA = 0.04292

UNITS: MM
DES: bhw/WA

30

Obj Height 77.8 mm
0.1 mm

Obj Height 54.5 mm
0.1 mm

AXIS
0.1 mm

ASTIGMATISM
S × T + (mm)

LONGITUDINAL
SPHERICAL ABER. (mm)

CHROMATIC
FOCAL SHIFT (mm)

DISTORTION (%)

LATERAL COLOR (mm)

UNITS: mm

FIELD: 77.8mm
IMAGE NA: 0.0429   EFL: 50mm
WAVELGTH: +:0.560 *:0.510 ◦:0.610 µm

DOUBLET MAGNIFIER (NEAR)
RAY TRACE ANALYSIS

bhw/WA
07 Jun 08
10:53 AM

Figure 8.2 5× doublet magnifier, lens data, lens picture, and aberration curves.

A real-world example will be helpful in demonstrating how a simple magnifier can actually improve our ability to visually resolve details. The back of a five-dollar bill contains a picture of the Lincoln Memorial. In the stone work above the twelve columns on the memorial, the names of several states have been engraved. Owing to the size (about 0.5 mm) and low contrast of the letters, the names of the states cannot be resolved with the naked eye. However, when this part of the image is examined using a 5× magnifier, the names of the states are easily readable.

The subject of trade-offs tends to come up frequently when discussing various optical situations. It will be noted that, when viewing with the naked eye, we can easily determine that it is a US five-dollar bill, containing numerous details including the picture of the Lincoln Memorial. When viewed using the 5× magnifier, we can now determine that the state name on the far left is Delaware, but the fact that this is indeed an entire piece of paper currency becomes a matter of speculation. In optical terms, the observed field of view has been reduced from around 140 deg with the naked eye to about 40 deg with the magnifier. We have traded a large field of view for finer detail within a smaller field.

A typical 5× loupe, or magnifier, will most often take the form of a cemented doublet, or a symmetrical Hastings cemented triplet lens. We will evaluate typical designs in both of these forms to identify any significant performance differences between the two. The 50-mm doublet that was shown in Fig. 8.1 is more completely described, and its performance characteristics are shown in Fig. 8.2. Similar information has been generated for a second 5× loupe in the form of a Hastings triplet (see Fig. 8.3).

In the lens pictures in both figures the final image surface is the convex curved surface on the far right. The field curvature shown there (about 0.5 D) has been introduced for evaluation purposes. The aberration curves shown indicate that, while both lenses are very well corrected on axis, in this particular case the doublet actually appears to offer a slight advantage in off-axis performance over the Hastings triplet.

While these aberration curves do give some indication of lens performance, a more useful tool would be one that yields an absolute value for lens resolution when used with the eye. This is best accomplished by computing MTF for each lens. The MTF curve plots the modulation (contrast) of the image formed by the lens as a function of image frequency. Figure 8.4 illustrates the concept of modulation and contrast, showing how they might be measured and calculated. Assuming a pattern of parallel black and white stripes with a range of frequencies, the modulation or contrast of the object (or image) is arrived at by comparing the maximum and minimum brightness values as follows:

$$\text{Modulation} = \left(B_{max} - B_{min} \right) / \left(B_{max} + B_{min} \right)$$
$$\text{Contrast} = \left(B_{max} - B_{min} \right) / B_{max}.$$

***PARAXIAL CONSTANTS**

Effective focal length:	50.05751	Lateral magnification:	0.18749
Numerical aperture:	0.04267	Gaussian image height:	15.00000
Working F-number:	11.71861	Petzval radius:	-65.47942

***LENS DATA**

TRIPLET MAGNIFIER (NEAR)

SRF	RADIUS	THICKNESS	APERTURE RADIUS	GLASS
OBJ	140.00000	-250.00000	80.00351	AIR
AST	--	25.00000	2.00000 AS	AIR
2	40.30000	4.50000	14.26743	F2
3	22.52000	13.65000	14.26743	BK7
4	-22.52000	4.50000	14.26743	F2
5	-40.30000	32.55000	14.26743	AIR
IMS	-- V	-- V	13.00000	

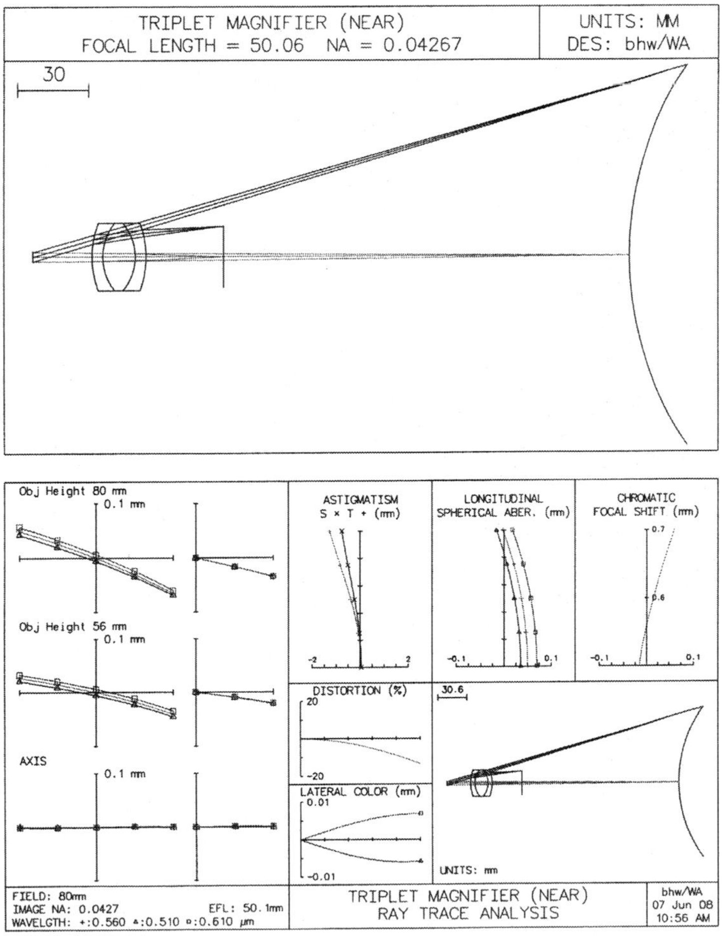

Figure 8.3 5× triplet magnifier, lens data, lens picture, and aberration curves.

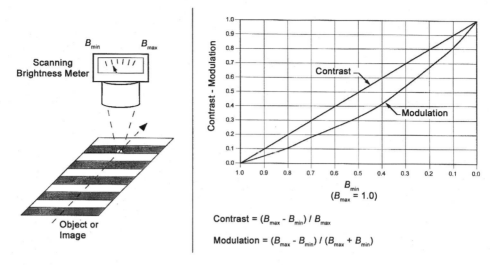

Figure 8.4 Image quality is often described in terms of contrast or modulation. This figure shows the definition of each and a typical method of measurement.

For MTF analysis, the object brightness from line to line is assumed to vary as a sine function. The modulation of a high contrast object is assumed to be 1.0 at all frequencies. At the image of that object, when the frequency is very low, the modulation will be close to 1.0. As the frequency at the image increases, the modulation of the image will decrease, i.e., the image will appear to be smeared as a result of diffraction effects and residual aberrations of the lens under test. Figure 8.5 shows the MTF curves for the two magnifier designs under consideration. That figure also contains a companion curve that is labeled "visual response curve." This curve indicates, for a typical visual system (eye) plus a 5.3× magnifier, the amount of image modulation that is required in order to visually resolve the corresponding frequency. While this typical visual response curve cannot be taken as absolute, it is valid for the relative comparison of two similar systems or designs. The point where this visual response curve crosses the MTF curve of the optics represents the maximum resolution possible when viewing a high contrast target with the 5.3× magnifier. Figure 8.5 shows that while both lenses are the same for on-axis imagery, for the off-axis case, the doublet will allow the viewer to resolve about 33 cycles/mm, while the triplet's off-axis resolution will fall to about 25 cycles/mm.

8.3 Eyepiece

The eyepiece is quite similar in function to the magnifier. It differs primarily in that the eyepiece is generally used in conjunction with other optics to form a complete instrument, such as a telescope or microscope. In that type of application, the eyepiece serves two functions simultaneously: first, it must project the final image to the viewer's eye, and second, it must form an image of

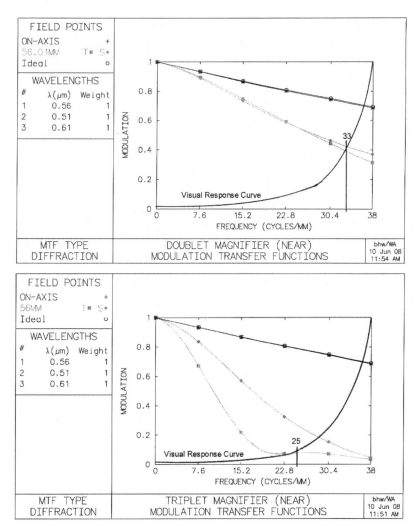

Figure 8.5 MTF data for the 50-mm (5.3×) doublet and Hastings triplet magnifiers.

the system aperture stop, which will be the exit pupil of that instrument. While the first function is by far the most important, proper eyepiece design requires that the second be paid some attention.

Eyepieces are generally more complex than the simple magnifier, and they become increasingly more complex as the field of view that must be covered is increased. That complexity is reflected in the number of elements that are required and the glass types that are used in the design. Figure 8.6 illustrates this point, presenting general information on six of the more common eyepiece designs. All of these eyepieces are drawn to the same scale and have the same 28-mm focal length. Detailed design and performance data on these and other eyepiece designs can be found in Chapter 5 of *Optical Design for Visual Systems*.[1]

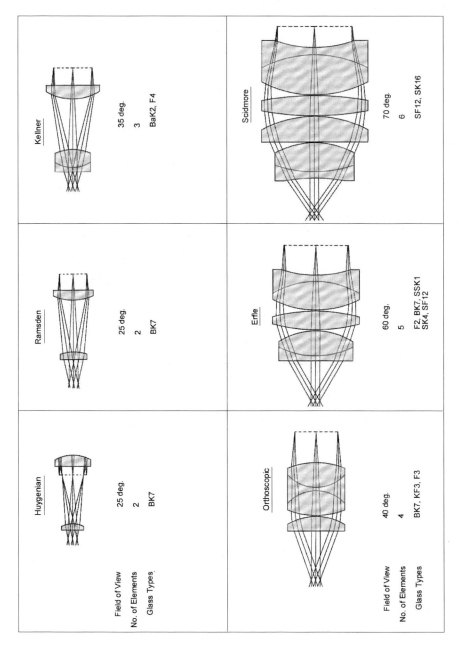

Figure 8.6 Family of common eyepiece designs, showing the increase in complexity as a function of the field of view that must be covered.

In most applications the eyepiece must be made axially adjustable to permit focus to accommodate for differences in the eyesight of viewers. A normal adjustment range of from +3 to –4 D will satisfy most requirements. The amount of eyepiece travel (in millimeters) can be calculated using the following formula:

$$\text{axial travel per diopter} = EFL^2 / 1000.$$

For a 28-mm EFL eyepiece, each diopter of adjustment will require:

$$\text{axial travel per diopter} = 28^2 / 1000 = 0.784 \text{ mm/D}.$$

Moving the eyepiece closer to the image being viewed will result in a diverging output beam, which corresponds to a negative diopter setting. To achieve the +3 to –4 diopter focus range, the 28-mm eyepiece will have to be moved from –3.2 to +2.4 mm relative to the infinity, or zero-diopter setting. The eyepiece focus design should provide a smooth, firm mechanism and an easily read scale indicating the approximate eyepiece diopter setting.

8.4 Microscope

A reasonable maximum limit for magnification by a simple magnifier would be about 20 to 25×. When a higher magnification is required, a compound instrument will be used. This compound magnifier is commonly referred to as a microscope. In its simplest form, the microscope consists of two lens assemblies: an objective lens and an eyepiece. The objective lens will form a magnified image of the object, while the eyepiece will be used to view that magnified image, thus providing additional magnification. Figure 8.7 shows a typical microscope arrangement, where the objective lens forms a 4× magnified image of the object, and the 10× eyepiece is used to view that image. The total microscope magnification is found by multiplying the objective magnification (4×) by the eyepiece magnification (10×), for a total magnification of 40×. Most often a microscope is designed by selecting from a wide variety of commercially available components. While custom microscope design is an active and challenging area of optical design, it will not be covered in this book.

The information presented in Fig. 8.7 represents a typical low-power microscope system constructed using commercial parts. The key to a successful design here is to identify a reliable source for quality parts and to work closely with the supplier to assure that all assumptions that might be made regarding optical and mechanical characteristics are valid.

Examining the details of Fig. 8.7, we see that both the objective lens and the eyepiece are similar designs, each made up of a pair of cemented doublets. For analysis purposes, the objective lens can be considered as imaging the object surface onto the field stop. The lens is designed to operate at a magnification of 4×. The field-stop size, a 16-mm diameter in this case, will limit the size of the object surface that will be seen. The objective lens assembly contains an aperture

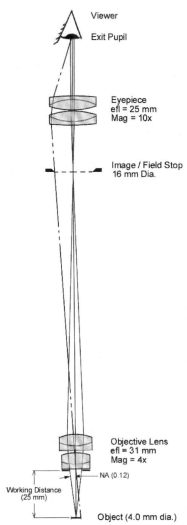

Figure 8.7 Typical compound microscope, constructed using standard, commercially available components.

stop which will limit the size of the on-axis light bundle. Vendor data indicates that the numerical aperture (NA) of this objective lens is 0.12. This means that the $f/\#$ of the used cone of light from the object will be:

$$ f/\# = \frac{1}{(2\mathrm{NA})} = \frac{1}{0.24} = f/4.17. $$

At the field stop this $f/\#$ will be increased by the magnification factor, making it $4 \times 4.17 = f/16.67$. Since we are using a $10\times$ (25-mm EFL) eyepiece, we can calculate the exit pupil diameter to be $25/16.67 = 1.50$ mm.

In a similar manner, we can analyze the microscope's field of view. The 16-mm-diameter field stop will limit the actual field of view to $16/4 = 4.0$-mm diameter at the object surface. When viewing the 16-mm-diameter field stop through the 25-mm EFL eyepiece, it will have an apparent angular half field of view whose tangent is $8.0/25.0 = 0.32$. From this we can see that the 4.0-mm-diameter object, which would subtend a half-field angle with a tangent of $2/250 = 0.008$ when viewed at a distance of 250 mm with the naked eye, will now subtend a half angle whose tangent is 0.32 when viewed through the microscope. This new apparent object size is magnified by a factor of $0.32/0.008 = 40\times$. This is consistent with the starting goal of a microscope having a magnification of $40\times$.

In the design of all optical instruments the subject of focus must be given some attention. In the case of the microscope shown in Fig. 8.7, the focus procedure would involve first moving the eyepiece to that point where the field stop is in sharp focus to the viewer's eye. After this has been accomplished, the entire microscope is moved to bring the object surface into sharp focus.

There are several ways by which a microscope can be used to make precise measurements of object details. The simplest and most obvious approach is to place the object on a movable platform (a stage) that is equipped with micrometers to measure its travel. The measurement is made by first aligning one extreme of the object being measured with a reticle pattern placed at the field-stop location. This zero-micrometer reading is recorded and the object is then moved to align its other extreme. The difference in micrometer readings will indicate the actual size of the object measured. Using this method, the overall accuracy is tied directly to the precision with which the two extreme settings have been made. Improved accuracy can be achieved by making the measurement on the image that is formed internally at the field-stop location. This method requires the use of a calibrated moving reticle and it is limited to measurements of objects no larger than about half the microscope's total field of view. Reticle calibration is accomplished by first viewing a precision standard scale (referred to as a *stage micrometer*) that is placed at the object surface. It will then be possible to establish the exact relationship between the amount of internal reticle travel and the reference scale. That relationship can then be applied to measurements of objects of unknown size.

Much has been done in recent years to combine the optics of the microscope with the modern compact digital camera. The camera may use an objective lens that will accept the output from the eyepiece of the microscope. In a preferred configuration, both the eyepiece and the camera objective can be eliminated and the camera sensor can be placed directly at the internal image plane of the microscope. This microscope-camera combination is especially effective in that the electronic camera data can be transmitted for remote viewing by a large number of observers.

8.5 Telescope

The telescope and the microscope have much in common, the major difference being that, unlike the microscope, the telescope is used to view large objects that are at great distances from us when it is not practical to reduce the distance to the object. We find that the origins of the prefixes *micro-* (small) and *tele-* (at a distance), are indicative of these conditions. We will discuss the design of two typical telescope applications, first for viewing the moon from our back yard, and then for looking at the batter in a ball game where our tickets are for the right field bleachers, some 400 feet from home plate.

In discussing the design of telescopes to perform these two functions we will hopefully learn much about the optical engineering that is involved. For observation of the moon (and other astronomical subjects) we might decide on a simple telescope made up of a refracting objective lens and an eyepiece. A practical limit on the size of an objective lens, based largely on cost, would be about 100-mm diameter, with a speed of $f/10$, making the focal length 1000 mm. For purposes of this design exercise, we will assume that we have located a reliable source for a family of good, reasonably priced, interchangeable eyepieces with focal lengths of 8, 15, and 28 mm, each designed to cover a 45-deg apparent field of view.

For our starting design exercise we will couple the 1000-mm objective lens with the 28-mm eyepiece. Telescope magnification is calculated by dividing the eyepiece focal length into the objective focal length. In this case the result will be 1000 mm/28 mm = 35.7×. The angular field of view that will be seen is limited by the diameter of the field stop in the eyepiece, which the supplier's data tells us will be 23.3 mm. The resulting field of view for the objective lens will be:

$$\tan^{-1}\left(23.3/1000\right) = \tan^{-1}0.0233 = 1.34\,\text{deg}.$$

Another important telescope characteristic is its exit pupil diameter, which will be equal to the objective lens diameter divided by the telescope magnification, in this case 100 mm/35.7× = 2.8 mm. These characteristics, as shown in Fig. 8.8, pretty much define the dimensional parameters of the telescope. Next, we will consider the optical performance, i.e., image quality limitations.

For an astronomical telescope objective lens it is reasonable to strive for near-diffraction-limited performance. In this case we can first apply the familiar formula to determine the radius of the diffraction-limited blur spot (the Airy disk):

$$\text{Airy disk radius } R = 1.22 \times \lambda \times f/\#,$$

where λ is the wavelength and $f/\#$ is the lens speed. For this objective lens,

$$R = 1.22 \times 0.00056 \times 10 = 0.0068\,\text{mm}.$$

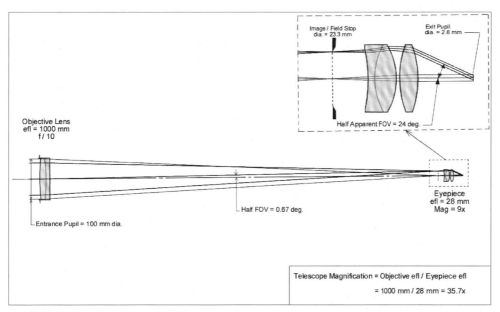

Figure 8.8 Basic optical parameters of an astronomical telescope made up of a 1000-mm, *f*/10 objective lens, and a 28-mm eyepiece.

Preliminary design analysis indicates that a well-corrected achromatic cemented doublet, made using ordinary optical glass types and optimized for visual use, will have a spot diagram, RED, and MTF as shown in Fig. 8.9. From the lens picture we conclude that lens shapes and thicknesses are reasonable from a manufacturing viewpoint. From the spot diagram we see that nearly all rays traced fall within the diameter of the Airy disk. From the RED plot we see that, with both geometric and diffraction effects included, about 80% of the on-axis light bundle will fall within the limits of the Airy disk. Finally, from the MTF plot we can see that the on-axis MTF curve is quite close to the ideal, or diffraction limit curve. In order to improve the design to perform at the diffraction limit it would be necessary to use very exotic and expensive optical glass types. The slight improvement in image quality would not justify the additional expense. The image formed by the objective lens will be viewed using the 28-mm EFL (9×) eyepiece. This eyepiece magnification will allow the average viewer to resolve a maximum frequency of 70 cycles/mm at that image. From the MTF curve in Fig. 8.9 it will be found that the diffraction effects of the *f*/10 objective lens will reduce this value to 65 cycles/mm, while the residual aberrations of the doublet will further reduce the maximum possible visual resolution to 63 cycles/mm. Each element (line or space) of this 63 cycle/mm pattern will be about 7.9-μm wide and will subtend an angle of 1.6 arcsec in object space. This 1.6-arcsec value represents the angular resolution limit for this telescope when it is used visually with a 28-mm-focal-length eyepiece. Another approach to determining the telescope's resolution limit would be to divide the visual limit (about 1 arcmin per element) by the telescope magnification (35.7×), which will yield the same 1.6-arcsec result.

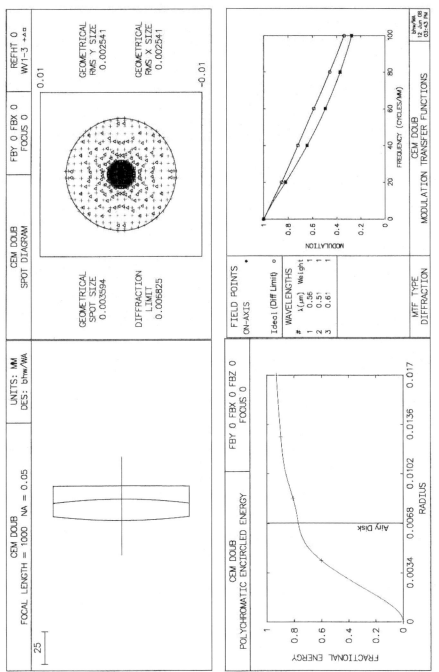

Figure 8.9 Lens picture and on-axis performance data for a 1000-mm, *f*/10 achromatic cemented doublet.

Figure 8.10 The field of view, as seen through a telescope with a 1.34-deg field of view (Fig. 8.8), observing the moon.

When the moon is viewed through this telescope, the field will appear as shown in Fig. 8.10. Looking through the eyepiece, the moon's disk will subtend an angle of about 20 deg to the eye. Without the telescope, the moon subtends an angle of 0.55 deg to the naked eye. Based on the visual resolution limit of 1 arcmin per element, we can conclude that, with the naked eye, we will resolve 0.55 deg × 60 = 33 elements across the moon's diameter (2160 miles). Thus, we can conclude that the smallest element on the moon's surface that we can resolve with our naked eye will be 2160 miles/33 = 65 miles. With the help of our new telescope, that number will be reduced to just under 2 miles.

Before we move on, this topic of maximum visual resolution deserves just a bit more attention. Certainly the naked eye can see a dark crater on the bright moon's surface (high contrast), even if that crater is less than 65 miles in diameter. But, if there are two 50-mile-diameter craters that are separated by 50 miles (center to center), the naked eye will not be able to determine that there are indeed two separate craters, i.e., they will not be resolved. However, if we look at these two craters with our new telescope, we will clearly resolve two separate and distinct craters. This would remain the case for a pair of 2-mile-diameter craters that are separated (again, center to center) by just 3 or 4 miles.

Additional telescope magnification will be introduced if we choose an eyepiece with a shorter focal length. We said earlier that we would choose interchangable eyepieces with focal lengths of 8, 15, and 28 mm. When used with the 1000-mm-focal-length objective these eyepieces will result in magnifications of 125×, 66.7×, and 35.7×, respectively. There are drawbacks to higher magnifications. First, any imperfections in the image formed by the objective lens duc to atmospheric disturbances, lens quality, telescope vibration, focus, etc., will be magnified by the shorter focal length eyepiece. Second, higher magnification will generally mean reduced field of view. In this case, the 1.34-deg field with the 28-mm eyepiece will be reduced to 0.7 deg when using the 15-mm eyepiece and to just 0.38 deg with the 8-mm eyepiece.

Another characteristic of the telescope that must be considered is the diameter and location of the exit pupil. Because we have an $f/10$ objective lens, it follows that the exit pupil diameter will be one-tenth the eyepiece focal length. That will be just 1.5-mm diameter for the 15-mm eyepiece and 0.8-mm diameter for the 8-mm eyepiece. An exit pupil diameter that is less than 1.0 mm will have a serious negative impact on the resolution capability of the eye. Finally, we should consider *eye relief*, which is the axial distance from the back of the eyepiece to the exit pupil. In order to see the entire apparent field of view, the observer must place his or her eye at a point where the eye pupil is in coincidence with the telescope's exit pupil. With conventional eyepieces the amount of eye relief will generally be about equal to the eyepiece focal length. Obviously, with just 8 or 10 mm of eye relief, comfortable viewing will be quite difficult for most, and impossible for viewers wearing eye glasses. For all of these reasons it will probably be best to use the 28-mm eyepiece for the majority of observations, the 15-mm eyepiece for somewhat increased magnification, and the 8-mm eyepiece only where very high magnification is essential. To summarize with a general statement, in cases like this the addition of higher magnification will often yield disappointing results.

One last word about our telescope . . . perhaps the ideal configuration for viewing and recording celestial objects today involves replacing the eyepiece with a modern digital camera designed specifically for astrophotography. Such a camera can be connected directly to a laptop computer where objects imaged by the objective lens can be viewed on the computer screen and then captured and stored for later examination and fine tuning.

The second telescope application that we will consider involves viewing a sporting event from a distance of some 400 ft. Why not use our astronomical telescope? There are a number of very good reasons, its size and weight for starters. The telescope will be about 5 in. in diameter, 40 in. long, and it will weigh at least 10 lbs. Its high magnification would require the use of a tripod and, if all that were not enough, the image that we would see would be upside down.

For comfortable hand-held viewing, the telescope magnification should not be greater than 10×. To provide a generous field of view to the eye, we will assume an apparent field at the eyepiece output of 50 deg. To assure good image brightness we will use an objective lens with an aperture (entrance pupil) diameter of 50 mm. Finally, to produce proper image orientation, we will assume the use of a roof Pechan prism in the back focal region of the objective lens. A telescope of this configuration, designed to view nearby objects, is referred to as a terrestrial telescope.

In order to keep the objective lens form simple it will be given a speed of *f*/8. This means the focal length of the objective will be equal to 8 × 50 = 400 mm. to achieve the desired 10× magnification, the eyepiece focal length will be 400/10 = 40 mm. The objective lens field of view will be 5 deg, which will result in an internal image size of 35-mm diameter. Knowing that the objective lens diameter is 50 mm and the image diameter is 35 mm, we can reasonably assume that the roof Pechan prism will require an aperture diameter of 40 mm. Figure 8.11 shows an optical layout of the resulting 10× telescope, suitable for viewing sporting events and similar terrestrial subjects. Telescopes of this type are generally specified in terms of magnification and aperture size. In this case, the telescope would be designated as 10 × 50, with a field of view of 260 ft at 1000 yds. Figure 8.12 illustrates the field of view that would be seen through this telescope. The advantage of using such an instrument can be appreciated when we consider the fact that, when viewed with the naked eye, the apparent size of the batter would be reduced by a factor of 10.

8.6 Binoculars

A pair of binoculars is made up of two identical terrestrial telescopes, linked together such that their optical axes are parallel. The distance between the two exit pupils is made to be adjustable to accommodate individual differences in eye separation (interpupillary distance, or IPD). The average value for IPD is about

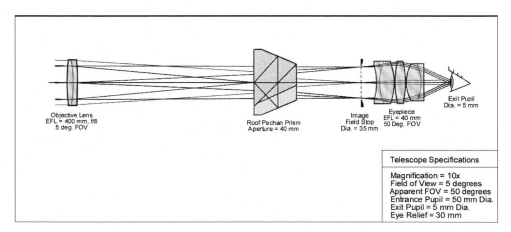

Figure 8.11 Basic optical parameters of a 10×, compact terrestrial telescope, consisting of a 400-mm, *f*/8 objective lens, a roof Pechan derotation prism, and a 40-mm EFL eyepiece.

Figure 8.12 The field of view, as seen through a 10× terrestrial telescope with a 5-deg field of view (Fig. 8.11), looking at a batter from a distance of 400 ft.

64 mm, with an adjustment range of +/–10 mm adequate to satisfy most requirements. It is critical that the two optical axes be parallel as they enter the viewer's eyes. A maximum misalignment tolerance of one arcminute will be accommodated by most users with little difficulty. The dual optical paths will result in more natural, relaxed viewing, with an enhanced stereo effect relative to that provided by a single telescope, i.e., a *monocular*. In general, the improvements in viewing provided by using binoculars versus a monocular of similar specifications and optical quality will be cosmetic in nature and, in many cases, will not justify the additional size, weight, cost, and complexity that would be involved. The optical layout of a typical pair of 7 × 50 Porro prism binoculars is shown in Fig. 8.13. Many recent binocular designs incorporate in-line erecting prisms which lead to a more compact, lightweight final design.

8.7 Riflescope

The *riflescope* is a low-power telescope designed specifically to improve the sighting accuracy of a rifle or similar weapon. There are several specific optical characteristics of the riflescope that do not apply to most other optical instruments. First, the required upright image orientation is accomplished using a relay lens rather than a prism assembly. While this leads to a longer overall

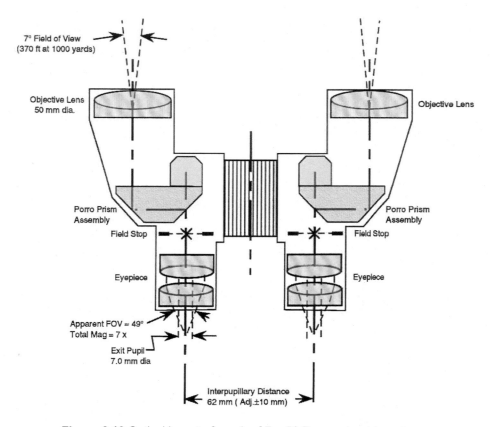

Figure 8.13 Optical layout of a pair of 7 × 50 Porro prism binoculars.

length, that length is compatible with the space available, and it actually enhances the ability to align the optical axis to the bore of the rifle. A reticle, capable of being adjusted in both azimuth and elevation, is an essential component of the riflescope in order to align the optical axis with the bore of the weapon. Riflescope optics must also provide generous eye relief to prevent impact with the users face when the rifle is fired.

From the layout shown in Fig. 8.14 it can be seen that the eyepiece contains the largest optical elements in the riflescope. The overall height of the instrument can be reduced if the vertical field of view is reduced. This is accomplished by placing a field stop at the eyepiece image, shaped as shown in the figure. This horizontally elongated field of view is consistent with the normal binocular visual field of view and it is also compatible with the function of the riflescope. Reduction of the vertical field of view allows the height of the larger eyepiece elements to be reduced proportionately. In many riflescope designs the relay optics can be adjusted, or zoomed, to produce a range of magnifications, which add greatly to the function of the instrument.

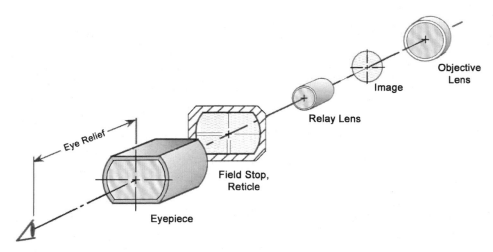

Figure 8.14 The optical system of the riflescope contains a relay lens to correct image orientation, and it requires greater-than-usual eye relief.

8.8 Surveying and Optical Tooling Instruments

Many optical instruments are used to make precision measurements of large objects at large distances. Depending on the application, this category of optical instrumentation has been given the designation of *surveying* or *optical tooling* instruments.

In the field of surveying instruments, the optical sight level and the transit (or theodolite) have been the two dominant optical instruments. The optical sight level is a precision high-power terrestrial telescope mounted in such a way that it will sweep a level plane, allowing the users to measure vertical departure from that plane. The transit and theodolite are also precision high-power terrestrial telescopes, mounted in such a way that their line of sight can be scanned in both azimuth and elevation. They are provided with precision scales that permit the precise measurement of angles between objects. The major differences between the transit and the theodolite are size (the theodolite is generally more compact) and the degree of precision with which angles can be measured (the theodolite is a more precise instrument). The optical configuration of the telescope used in both of these instruments will usually consist of an objective lens, a focusing relay lens, a reticle, and an erecting eyepiece assembly that resembles a low-power microscope in both its construction and its function (see Fig. 8.15).

The field of optical tooling uses surveying instruments as well as a family of special optical tooling instruments that includes collimators, autocollimators, alignment telescopes, and a variety of special-purpose accessories. The collimator contains an illuminated reticle set precisely at the focal plane of the objective lens. The projected (collimated) reticle pattern establishes a reference line of sight to which any number of other optical tooling devices can be aligned, thus establishing their relationship to one another.

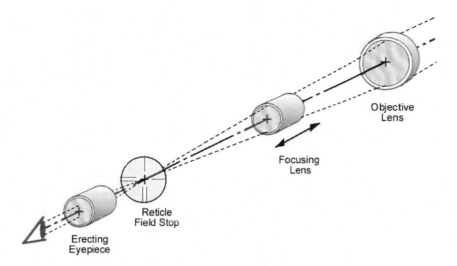

Figure 8.15 Optical layout of a typical telescope system for use in a sight level or a transit, two of the more common optical instruments used for surveying.

The autocollimator is similar to the collimator, but it is equipped with an eyepiece and a beamsplitter for purposes of simultaneous viewing and reticle projection. The process of autocollimation involves projecting the reticle such that it reflects from a flat reference mirror. That reflected reticle pattern is then viewed through the eyepiece of the autocollimator. The reflected pattern will be in precise alignment with the reticle at the eyepiece image plane only when the mirror surface is precisely perpendicular to the line of sight. Here is another way of describing the process of autocollimation: an image of the autocollimator is formed behind the reference mirror. In order to sight squarely into that image, the mirror must be set precisely at a right angle to the optical axis of the autocollimator. Figure 8.16 illustrates the basic optical layout of the collimator and autocollimator.

The alignment telescope [also called a jig alignment telescope (JAT)] is an autocollimator with the additional capability of making precise linear measurements in the *X-Y* plane at relatively great distances. This is accomplished by including the capability to focus on objects at all distances, along with an optical micrometer, which permits precise, calibrated displacement of the JAT's projected line of sight. In its most basic form the optical micrometer is a plano-parallel window, placed ahead of the objective lens and capable of being tilted through a calibrated angle, thus displacing the line of sight through a precisely proportional distance. A micrometer drum is attached to the tilting plate, giving a direct readout of the amount of displacement. Figure 8.17 illustrates the optical micrometer and includes formulas for line of sight displacement as a function of plate tilt angle.

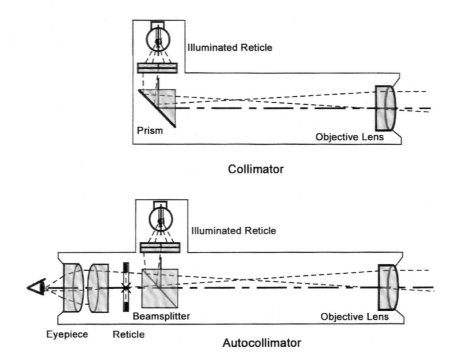

Figure 8.16 Collimator and autocollimator, two basic optical instruments used in optical tooling applications.

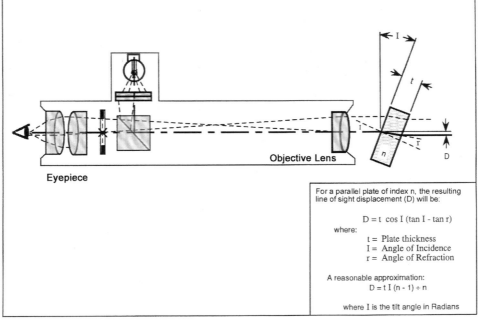

For a parallel plate of index n, the resulting line of sight displacement (D) will be:

$$D = t \ \cos I \ (\tan I - \tan r)$$

where:

t = Plate thickness
I = Angle of Incidence
r = Angle of Refraction

A reasonable approximation:
$$D = t \ I \ (n - 1) \div n$$

where I is the tilt angle in Radians

Figure 8.17 The optical micrometer is a plano-parallel plate which, when titled as shown, will precisely displace the line of sight by an amount *D*.

8.9 Periscope, Borescope, and Endoscope

This family of optical instruments is distinguished by the fact that its overall length is many times greater (often more than one hundred times greater) than its diameter. The periscope may range in complexity from the cardboard tube with flat mirrors at opposite ends that we typically see at the golf course, to the modern submarine periscope that may be 7 in. in diameter, over 40 ft in length, and may contain more than 100 optical elements (see Fig. 8.18). The "golf course" periscope does nothing to enhance our visual capability except to displace the line of sight vertically, so that one can see over the heads of others in the crowd. The submarine periscope, on the other hand, can provide several levels of optical magnification and a family of sensors to greatly improve the viewer's ability to observe and record the scene being viewed.

The *borescope*, while similar in optical design to the submarine periscope, differs greatly in that it is much smaller. A typical borescope might be just 8 mm in diameter and 500 mm in length. Its name is derived from early applications where it was used to inspect the inside diameter (bore) of a rifle or other gun barrel. The *endoscope* is a borescope that has been specifically designed to penetrate and view inside the body. Endoscopes designed for specific medical applications may take the name of the procedure or part of the body involved (e.g., arthroscope for joints, laparoscope for the abdominal cavity).

Common to this family of instruments is an optical layout consisting of a high-resolution objective lens, a reticle, several relay lenses, and a camera or an eyepiece. Figure 8.18 shows the optical configuration of a modern submarine periscope with both visual and camera capability. Early periscope designs were accomplished with a single set of relay lenses in the mast optics. That led to periscope designs where the vignetting of off-axis light bundles was so great that the image brightness at the edge of the field of view might fall to just 10% of the on-axis value. The versatility of the eye was such that this did not represent a serious problem. However, during World War II (early 1940s) the navy started to use periscopes for photographic missions and the poor relative illumination led to poor photographs. The solution was to add two more sets of relays (one additional set would have produced an inverted image) to the optical design. In this configuration the relative illumination was increased to 40%, a variation that could be handled by the exposure tolerance (latitude) of most cameras. Another popular function commonly incorporated into the submarine periscope involved an optical means of estimating the range to the target. This usually required knowledge of the target size and a means of measuring image size. In recent years this range-finding process has become simpler and more accurate with the incorporation of the laser range finder into the periscope's optical system.

The borescope is quite similar in theory to the submarine periscope, with the major exception of its size. While the diameter of the borescope may typically range from 4 to 12 mm, the lenses that make up the borescope optical system may be as small as 1 mm in diameter. Figure 8.19 shows a typical optics layout

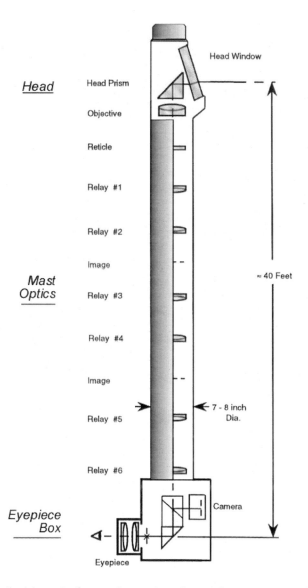

Figure 8.18 Optical layout of a modern submarine periscope system. Elevation scan is accomplished by rotation of the head prism, while the viewer rotates the entire periscope about its vertical axis for azimuth scan.

for a borescope. Along with the viewing optics, the borescope shown also contains a coaxial bundle of fiber optics to deliver illumination to the field being viewed. It should be noted in the figure that the lenses in the borescope do not conform to the usual proportions of length to diameter (typically 1:8). Because these optics are so small in diameter, their relative thickness must be increased substantially to facilitate manufacture and to maintain alignment within the final assembly. While the straight-ahead viewing shown is a popular configuration, it is possible to add a head prism or mirror to permit viewing at any desired angle.

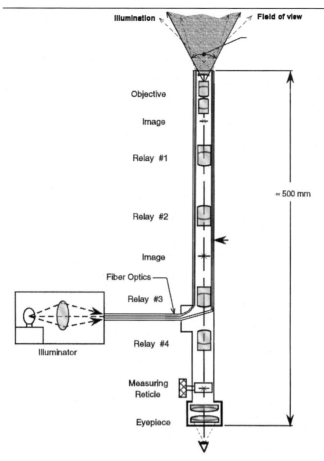

Figure 8.19 Optical layout of a typical borescope system. Unique to the borescope is the requirement that it contain an integrated light source to illuminate the object being viewed.

The rapidly developing field of compact solid-state electronics for imaging has caused several major changes in the field of periscopes and borescopes. The developing trend for all of these instruments is to place the detector as close as possible to the objective lens. This often eliminates the need for relay optics and will, in many cases, result in improved image quality. Once the image has been formed and converted to an electronic signal by the camera, it can be transmitted via wire, or optical fiber, to a remote location for viewing, broadcast, or recording. In the submarine periscope version of this configuration, a compact optical system is used in conjunction with high-resolution monochrome and color TV cameras in the head of the periscope. Camera output is transmitted to the ship's command center, where it is displayed on a bank of high-resolution monitors.

In the case of the borescope for medical applications, recent designs have been accomplished using what is referred to as the "chip in the tip" approach. Here the image is formed using a high-precision, micro-objective lens onto a solid-state video chip that may be as small as 1.5 mm^2. Output from the chip is then transmitted to auxiliary electronics outside of the borescope where it is converted to a conventional TV signal for viewing by the medical team. Several design configurations have been developed recently that permit the forming of dual images which, when viewed under the proper circumstances, present a true three-dimensional stereo image to the observer. The advantage of this approach to certain medical procedures has been significant.

In certain applications it is advantageous to have a flexible borescope rather than a rigid assembly. This flexibility has been accomplished using a flexible coherent fiber-optics bundle to transfer the image from the tip of the instrument to the output end. This approach will be limited by the resolution of the fiber bundle and its light transmission efficiency. More recently, the electronic approach has made it possible to form the image in the tip and then transfer it electronically, through a flexible cable, to an external point for further processing.

While many of these recent developments have led to optically simpler systems with reduced numbers of lenses, the compact nature and precision requirements of the resulting optical systems have challenged, and will continue to challenge, the skills and expertise of the optical engineer and lens designer.

8.10 Review and Summary

This chapter has presented a review of the field of basic optical instruments. The primary intent has been to familiarize the reader with some of the optical engineering considerations that are involved in the design of such instruments. From the simplest magnifier to the optically complex submarine periscope, it is the optical designer's responsibility to understand the intended function of the instrument and to ensure that the optical system is capable of carrying out that function. Most important is the final evaluation of the instrument's image quality. In many cases the eye is the system detector, and as such, its capabilities must be included in any performance evaluation. Several examples are presented throughout this book demonstrating how that can be done. In a subsequent chapter more detailed information on the characteristics and performance of the human visual system will be presented.

In several instances this chapter has indicated the impact that modern technological developments, particularly in the area of electronics and detectors, have had on instrument design. The laser range finder and compact solid-state video and still cameras are examples of components that have produced significant changes throughout this field in recent years. The lesson offered here is that it is extremely important that the optical engineer remain current on technological developments and be prepared to explore methods of including them in any instrument design.

Reference

1. B. H. Walker, *Optical Design for Visual Systems*, SPIE Press, Bellingham, Washington (2000).

Chapter 9
Optical Materials and Coatings

9.1 Introduction

This chapter will be devoted primarily to discussing a variety of materials that are used in the design and manufacture of optical components. The most common of these is optical glass, in its many forms. Other materials, such as plastic and fused silica, are sometimes substituted for optical glass in order to take advantage of their special characteristics. Optical coatings are added to components to enhance their optical performance. Antireflection coatings will improve the light transmission of a lens. Single- and multiple-layer antireflection coatings will be covered. Materials for use in infrared systems are unusual in several respects. The basic characteristics of the most commonly used IR materials will also be presented.

All these topics will be covered in an introductory fashion, with the intent to provide the reader with a level of knowledge that will permit you, as the optical designer, to generate basic specifications and discuss requirements with specialists in these fields.

9.2 Optical Glass

The material we refer to as *optical glass* differs from the more common glasses in that its make up and ultimate performance characteristics are precisely controlled during all stages of the manufacturing process. This permits the optical designer to generate a design configuration based on published catalog data that, when manufactured, will perform exactly as indicated by the design data.

There are several reliable, established sources for optical glass throughout the world. Primary sources manufacture a complete variety of precision optical glasses from basic raw materials. Other glass suppliers act as processors, taking bulk supplies of glass from the original manufacturers and modifying the size and shape of starting glass blanks, and precisely controlling the characteristics of the glass during this process. It is enthusiastically recommended that the interested reader visit available web sites, contact all glass manufacturers and suppliers, and request catalog and technical data describing their products and services (see

Appendix C). With this information, it will then be a simple matter to select the most appropriate source to meet your requirements.

Much of the information that follows will refer to data and materials supplied by the Schott Optical Glass Company of Duryea, Pennsylvania and Schott Glaswerk of Mainz, Germany. I have opted to present this information because Schott is the glass company with which I have had the most personal experience and, as a result, have collected the most data and general information. This choice on my part is not intended to indicate a superiority of Schott Glass over the many other established suppliers of optical glass.

Optical glass must have several unique characteristics in order to perform its required function. First, it must be transparent to the wavelengths of the design, and its index of refraction over the spectral band of the design must be known with a high degree of precision. Beyond these critical factors, it is important to take into account such factors as the cost and workability of the glass, along with its many physical, mechanical, and thermal characteristics.

Individual optical glass types are identified by both name and number. The name is assigned by the manufacturer and will generally identify the glass type based on the key elements that it contains. For example, typical Schott glass names are BK (borosilicate crown) and LaF (lanthanum flint). Other glass producers have developed their own methods for naming glasses. As a result, the same glass type may be known by several different names, depending on the source.

The glass identification number is a six-digit code derived from the index of refraction and dispersion of the glass. If, for example, the index of a glass is 1.517 and the dispersion is represented by an Abbe v number of 64.2, the numerical identifying code for that glass would be 517642. The glass identification number has the obvious advantage that it is not optional on the part of the glass manufacturer, but is tied directly to the optical characteristics of the glass, regardless of the source.

The glass map shown in Fig. 9.1 indicates the index of refraction and the Abbe v-number for today's most commonly used optical glass types. The Abbe v-number is an indicator of the dispersion characteristic of the glass. The v-number is calculated using the following formula:

$$v_d = \frac{n_d - 1.0}{n_f - n_c},$$

where n_d, n_f, and n_c are the indices at wavelengths 0.588 μm (yellow), 0.486 μm (blue), and 0.656 μm (red), respectively.

The v_d number for an optical glass indicates the amount of dispersion that will occur across the visible portion of the spectrum, with its numeric value inversely proportional to the amount of dispersion that will actually occur. As Fig. 9.1 shows, the v number for the majority of glasses will range from 25 to 65,

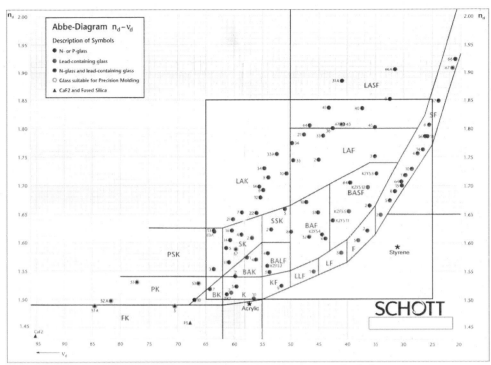

Figure 9.1 Typical glass map showing common optical glass types.

with common values for crown glass at about 60 (less dispersion) and flint glass about 30 (more dispersion). It will also be seen from the glass map that the index of refraction for these most frequently used glass types will range from 1.50 to 1.85.

The significance of the index of refraction change with wavelength (dispersion) is shown in Fig. 9.2. For this example, data for two simple lenses—one made from common crown glass (BK7) and the other from a common flint glass (SF1)—are shown. Both lenses have a nominal focal length of 100 at a wavelength of 0.588 µm. The curves show how the focal lengths of these lenses will change when they are used at other wavelengths. Over the wavelength range from 0.35 to 1.1 µm, the focal length of the crown lens will vary from –4 to +2%, while the flint lens focal length will change from –10 to +4%.

In order to generate a proper and functional optical design, it is essential that many other characteristics (beyond index and dispersion) of the glasses be carefully considered. The internal light transmittance of most optical glasses will be quite high for wavelengths from 0.4 to 2.0 µm (see Fig. 9.3). When system transmission is critical near either extreme of wavelength, more detailed information on the particular glass types being considered must be referenced. There is a wealth of additional information available from the glass manufacturers detailing the optical, mechanical, and thermal characteristics of its glasses. The optical designer should carefully review all of this information and

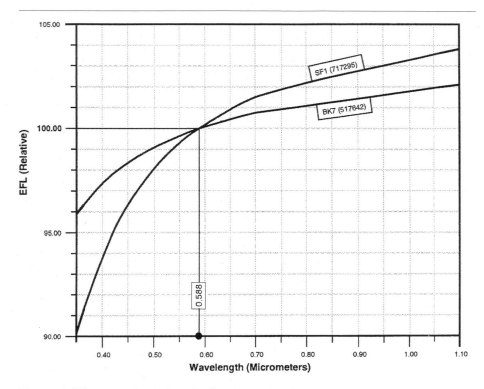

Figure 9.2 Because the index of refraction varies with wavelength, the focal length (EFL) of a simple lens will also vary, depending on wavelength. Curves shown here indicate how the focal length will vary for a 100-mm EFL lens made from common crown (BK7) and flint (SF1) optical glasses.

consider it with respect to the detailed requirements of the particular system being designed.

Catalog data will indicate the nominal index of refraction for each glass at a number of wavelengths. Normal procurement will result in glass that is within ±0.001 of these catalog index values, with a ±0.8% tolerance on the v number. The designer must determine if this level of index control is consistent with system requirements. When required, the index can be controlled to as close as ±0.0002, with a corresponding v number tolerance to ±0.2%. These reduced tolerances will result in a slight increase in the cost of the glass. When ordering glass, it is usually possible to obtain melt data from the glass supplier. This melt data will include precise, measured, index-of-refraction data for the particular glass blanks that have been delivered. With this melt data in hand the designer can actually simulate the final design on the computer and fine tune the lens shape if required.

In addition to the basic index value for any glass blank, it is equally important to control the homogeneity throughout that blank. Homogeneity is the amount of variation in index of refraction that is allowed throughout the volume of any one glass blank. The standard (default) value for homogeneity within a

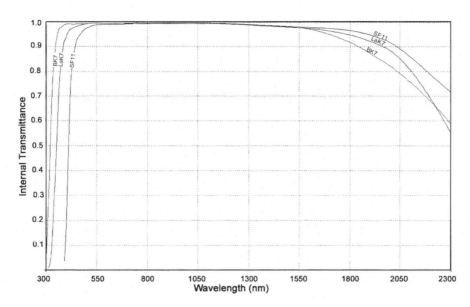

Figure 9.3 Internal transmittance for 25-mm-thick samples of three common optical glass types.

melt is 1×10^{-4}. The homogeneity of any individual piece taken from that melt will be less. When required, it is possible to produce glass blanks with homogeneity values as small as 5×10^{-7}. Homogeneity becomes particularly important in optical components that are to be used in a near-diffraction-limited system.

Bubbles, *seeds*, and *striae* are categories of internal glass defects that may be observed, usually under very sensitive viewing conditions. Bubbles are formed when small air pockets are left as the glass cools. Seeds and striae are solid contaminants within the glass: seeds are near-spherical, while striae are long and string-like. Except for a few very unusual cases, most of today's optical glasses will not contain bubbles, seeds, or striae to any significant degree. Optics designed to transmit high-power lasers may require special consideration regarding bubble and seed content.

The thermal characteristics of the glass must be considered when the temperature range over which the lens is to be used (or stored) is substantial. There are two values that enter into thermal considerations. They are the coefficient of linear expansion α and the change in index with temperature dn/dt. The linear expansion of an optical element will most generally be acceptable if that element has not been mechanically constrained such that severe stress or possible breakage of the element might occur as a result of temperature changes. The change in index with temperature dn/dt will produce a focus shift and possible significant degradation of image quality if refocus is not possible. This change in focus can be analyzed and, in some cases, compensated for by the proper choice of lens cell material. This area of system design is a joint responsibility to be shared by the optical and the mechanical design engineers.

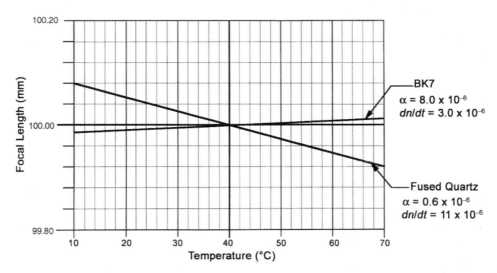

Figure 9.4 The focal length of a lens will vary as the ambient temperature changes. Two factors are involved: the linear expansion α of the lens material and the change in index of refraction with temperature (*dn/dt*).

9.3 Low-Expansion Materials

In a number of applications it is desirable to use optical materials that have a low coefficient of thermal expansion. Most common of these is the case of the first-surface mirror blank. One economical choice in this application is the material produced by Corning, with the trade name *Pyrex*. This is a very stable optical material with a thermal expansion that is less than half that of typical optical glasses. Pyrex is not suitable for most lens applications due to internal coloration and defects, such as bubbles and seeds.

Lenses and higher-quality mirrors can be made from *fused quartz* (also called fused silica). Fused quartz is optically pure and has a coefficient of expansion that is less than one-tenth that of typical glass. When even greater thermal stability is desired, there are several materials with expansion coefficients that are essentially zero (Schott Zerodur, Corning ULE, and Owens CERVIT). As with Pyrex, these materials are most suitable for use as mirror substrates, not as transmitting lens blanks.

It is interesting to note that, while fused quartz is an ideal lens blank material in terms of its optical quality and low expansion coefficient, it does have a relatively high *dn/dt* characteristic. The result is that, in a lens application, while the size and shape of the lens will remain essentially constant with change in temperature, the focal length of the lens will fluctuate due to the change in index of refraction. The data shown in Fig. 9.4 illustrates the change in focal length with temperature for a lens element made from fused quartz as compared with a similar lens made from BK7 optical glass. In the case of the fused-quartz lens the *dn/dt* is essentially the sole cause of the focal length change, while for the BK7 element the linear expansion of the lens, which increases its EFL, is partially offset by an increase in the index of refraction, resulting in a net focal length change that is considerably less than that of the fused-quartz lens. This is a very basic demonstration of how a change in temperature can affect the focal length of a lens element. Any optical system that is subject to substantial temperature fluctuations must be carefully analyzed, taking into account the expansion coefficients of all optical and mechanical parts, combined with the *dn/dt* of all refracting optical materials. Most of today's computer programs for optical design include routines written to perform this type of analysis.

9.4 Surface Losses and Antireflection Coatings

When light passes into or out of an optical component there will be a reflection loss at the air-to-glass interface. The amount of that loss will be proportional to the index of refraction of the glass from which the component is made. The general formula used to compute the surface reflection loss (air to glass) is as follows:

$$R = \frac{(n-1)^2}{(n+1)^2}.$$

For a standard optical glass with an index of refraction of 1.5, this formula indicates a loss of about 4%. For glass with an index of 1.8, the surface reflection loss will reach 8%. When several surfaces are involved it is important to remember that the absolute loss per surface becomes slightly less at each surface. Assume that we have five elements made from glass with an index of 1.8. The final transmission, ignoring absorption losses, through the 10 air-to-glass surfaces would be found using the following formula:

$$t = 0.92^{10} = 0.43, \text{ or } 43\%.$$

As lens designs have become more complex, involving the use of more elements and high-index glasses, the need to reduce surface losses has become critical. It was discovered that if a very thin layer of the properly chosen transparent material was added to the glass surface, those losses could be reduced substantially. The key to a successful antireflection (AR) coating was to select a material with an index of refraction that was equal to the square root of the glass index, and to control the thickness of the layer to be exactly one-quarter of the wavelength being transmitted. Figure 9.5 shows the transmission of a single optical element, uncoated and coated with a single layer of AR coating, as a function of the element's index of refraction.

It becomes clear that rarely can either or both of these criteria be met exactly. The selection of available coating materials is quite limited, making the index of refraction match quite difficult. The one-quarter-wave thickness is not a problem for a monochromatic system. However, for the more common system, designed

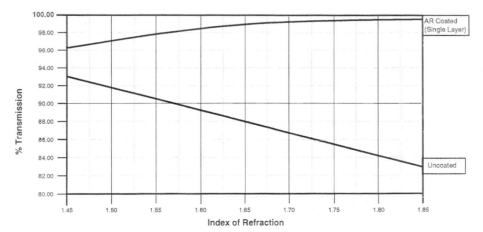

Figure 9.5 Transmission curves through a glass window as a function of the index of refraction. The lower curve is for an uncoated window, the upper curve is for a window with a single layer of magnesium fluoride antireflection coating on both sides.

to cover a broad wavelength range, a compromise thickness must be selected, usually based on the center of that range. The development of multilayer coatings has made it possible to accomplish most reasonable transmission goals with just a slight addition of cost and complexity to the final lens assembly. The data shown in Fig. 9.6 illustrates the effectiveness of single and multilayer AR coatings on

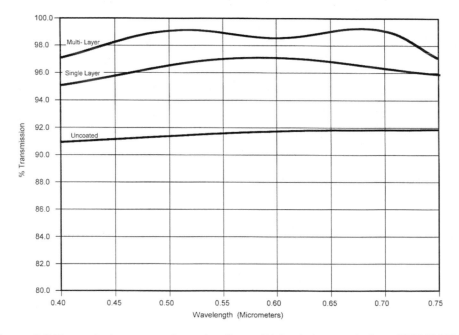

Figure 9.6 Transmission curves through a 5-mm thick window made from BK7 (517642) optical glass, uncoated and coated on both sides with single- and multiple-layer AR coatings.

the transmission of typical lens elements with both surfaces coated. It will be noted that for glass of normal index (about 1.5), the improvement in transmission compared to an uncoated element is about 5% when a single-layer coating is used, with an additional 2% gain possible when a multilayer coating is used. In general, for systems operating over the visible spectrum, single-layer coatings are most cost effective on high-index glasses, while for glasses in the normal index range, multilayer AR coatings are usually the best choice.

9.5 Materials for Infrared Systems

In recent years many applications have been found for optical systems designed to operate in the infrared portion of the spectrum. Such systems are designed to sense thermal radiation emitted by the object being viewed. As such, it is possible to view objects that are not illuminated and may not be visible in the traditional sense. The most common applications of IR systems involve night vision. Most of the useful IR energy emitted by warm or hot objects will have wavelengths in the 3–12-μm spectral range. For systems that are operating in the earth's atmosphere, much of this energy will not be transmitted. As Fig. 9.7 shows, the atmosphere transmits two distinct IR bands, one from 3 to 5 μm and the second from 8 to 12 μm. This section will deal with four of the most popular materials used in IR optical system designs covering these spectral bands. They are germanium, silicon, zinc sulfide, and zinc selenide.

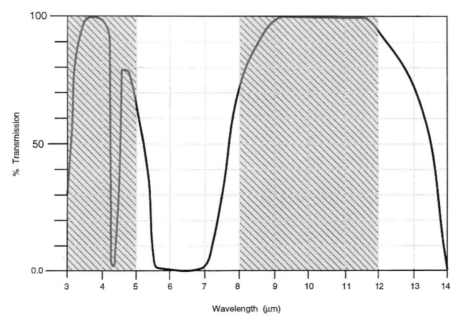

Figure 9.7 Smoothed curve, showing the two IR windows (3–5-μm and 8–13-μm) created by the average transmission of the earth's atmosphere (approximate path length of 0.5 km at sea level).

9.5.1 Germanium

Germanium (Ge) is a popular material for lenses designed to transmit IR energy in either the 3–5-μm or 8–12-μm spectral bands. Germanium has an index of refraction of approximately 4.0 at these wavelengths. Because of this high index, the surface reflection losses for a single uncoated air-germanium surface will be so great (≈36%) as to make the use of an uncoated germanium optic totally impractical. A germanium window that is uncoated will transmit only 47%, including all internal reflections. Broadband AR coatings for germanium will improve the transmission of a single element to around 93% over the 3- to 12-μm range. Coatings designed for either the 3–5-μm or 8–12-μm spectral bands will increase element transmission to about 98%.

There are three popular AR coating designs used on germanium optics, the selection of which depends on the environment to which the component will be exposed. Figure 9.8 shows the transmission curves for a single germanium element coated with a very durable single layer (diamond coat), a durable multilayer coating, and a less durable multilayer coating suitable for components that are protected from harsh environmental conditions and not subjected to frequent cleaning.

Germanium has mechanical characteristics and appearance that are more like metal than glass. With a density of 5.32 g/cm^3, germanium weighs about twice as much as glass. Its metallic characteristics make germanium a good candidate for the micromachining process. This makes the production of aspheric surfaces on germanium lenses a relatively straightforward matter. This feature, combined with its high index of refraction, leads to many unusually simple but effective optical designs for IR systems using germanium.

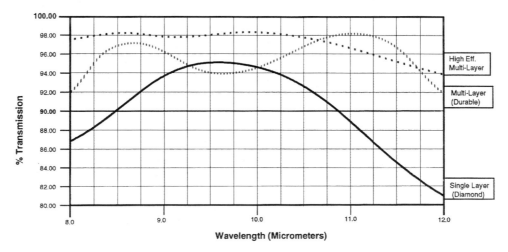

Figure 9.8 Transmission curves through a germanium window with a single-layer AR coat (diamond), durable multilayer, and (less durable) high-efficiency multilayer AR coatings on both sides.

9.5.2 Silicon

Silicon (Si) is a popular material for lenses designed to transmit IR energy in the 3–5-μm spectral band. Silicon has an index of refraction of approximately 3.43 at these wavelengths. Because of this high index, the surface reflection losses for a single uncoated air-to-silicon surface will be so great (≈30%) as to make the use of an uncoated silicon optic totally impractical. A silicon window that is uncoated will transmit about 54%, including all internal losses. A single-layer AR coating used on both surfaces will improve the transmission of a silicon element to around 90% in the 3--μm spectral band. With a density of 2.33 g/cm³, silicon weighs about the same as glass. Silicon lenses can be produced using the micromachining process, making the production of aspheric surfaces a relatively simple matter.

9.5.3 Zinc sulfide

Zinc sulfide (ZnS) is a popular material for lenses designed to transmit IR energy in either the 3–5-μm or 8–12-μm spectral bands. Zinc sulfide has an index of refraction of approximately 2.2 at these wavelengths. Because of this high index, the surface reflection losses for an uncoated air to ZnS surface will be so great (≈14%) as to make the use of an uncoated zinc sulfide element impractical. A zinc sulfide window that is uncoated will transmit about 75%, including all internal losses. A broadband AR coating for zinc sulfide will improve the transmission of a single element to around 93% over the entire 3– 12-μm spectral band. Coatings designed specifically for either the 3–5-μm or 8–12-μm spectral band will improve single-element transmission to about 98%. Zinc sulfide has mechanical characteristics and appearance that are similar to optical glass. With a density of 4.1 g/cm³, zinc sulfide weighs nearly twice as much as the average glass. The internal transmission of normal zinc sulfide is good over the spectral band from 3 to 12 μm. In addition, it is possible to treat normal zinc sulfide in manufacture such that its transmission window is increased to include most of the visible portion of the spectrum. The resulting material is known as multispectral-grade zinc sulfide and also by the trade name Cleartran (see Fig. 9.9). This improved visible transmission is very helpful in the alignment of optical systems with lenses made from zinc sulfide.

9.5.4 Zinc selenide

Zinc selenide (ZnSe) is another popular material for lenses designed to transmit IR energy in either the 3–5-μm or the 8–12-μm spectral bands. Zinc selenide has an index of refraction of approximately 2.4 at these wavelengths. Because of this high index, the surface reflection losses for an uncoated air-zinc sulfide surface will be so great (≈17%) as to make the use of an uncoated zinc selenide element impractical. A zinc selenide window that is uncoated will transmit about 71%, including all internal losses. Antireflection coatings for zinc selenide will improve the transmission of a single element to around 93% over the broadband

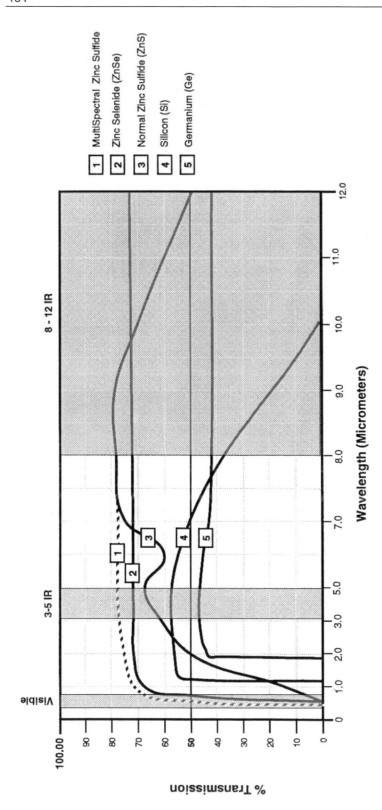

Figure 9.9 Transmission curves for four of the most common refracting IR materials. Special treatment of zinc sulfide results in increased visible transmission for the multispectral grade (also known as Cleartran). In regions of maximum transmission, losses are due primarily to surface reflections.

from 3 to 12 µm. Other coating designs will bring the transmission to 98% in either the 3–5-µm or the 8–12-µm spectral bands. Zinc selenide has mechanical characteristics and appearance that are similar to optical glass. With a density of 5.27 g/cm^3, zinc selenide weighs nearly twice as much as average optical glass. The transmission of zinc selenide covers the spectral band from 0.6 to 20 µm. While appearing slightly yellowish in color, zinc selenide does transmit enough visible energy to permit relatively simple alignment of optical elements. Zinc selenide is noted for its superior optical properties, which include very low absorption at the infrared wavelengths.

Figure 9.9 contains the basic transmission curves for all of these IR materials. Note that these curves are for uncoated elements, which indicates the substantial surface reflection losses that will occur.

9.6 Optical Plastics

Optical-quality plastics bring several distinct advantages to the design process. Plastics are very strong, capable of surviving conditions of severe shock that would destroy similar elements made from optical glass. Plastics are also light in weight, nominally one-third the weight of the average optical glass. While many plastic optical components are produced by the molding process, plastic can also be worked using conventional machining steps with precision tools, followed by modified optical polishing techniques. The decision regarding the manufacturing process to be used will be based on several factors, including the optical quality required, the quantity to be produced, and the size of the components.

As usual, there exists a set of tradeoffs to counter these many advantages. Optical plastics are quite difficult to finish to standard optical tolerances. Once finished, the optical stability of the plastic component is less than that of a comparable glass part. Figure 9.10 illustrates the change in focal length that will occur in a plastic lens as the temperature changes. Obviously, some method of maintaining focus during temperature shift (athermalization) must be part of any lens design using plastics. Plastic is relatively soft, which often leads to scratched optical surfaces during production and when parts are being cleaned.

The variety of optical plastics available is quite limited. The vast majority of plastic optics are made from polymethylmethacrylate, or acrylic. Acrylic is favored for its scratch resistance, optical clarity, and mechanical stability. With an index n_d of 1.491 and a v number of 57.2, acrylic may be thought of as the crown glass of plastics. Second in popularity among the optical plastics is polystyrene (styrene). With an index n_d of 1.590 and a v number of 30.8, styrene is a logical choice to act as a flint glass substitute in many plastic lens designs. There are just a few other plastic materials available for special applications, but in most cases acrylic and styrene are the materials best suited. This limits the variety of lens designs that can be generated using plastics. Acrylic (491572) and styrene (590308) optical plastics have been added to the glass map shown in Fig. 9.1.

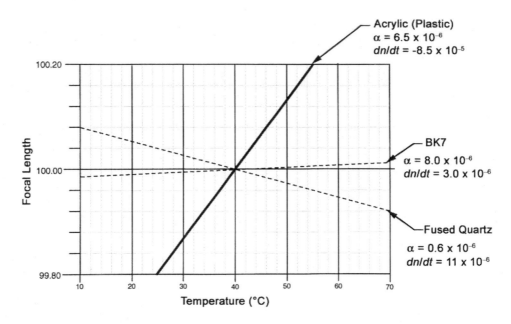

Figure 9.10 The focal length of an optical element will change as the ambient temperature changes. In the case of acrylic (plastic), the rate of change is affected by large thermal constants (α and *dn/dt*) that add rather than cancel.

In summary, it is essential that the application be carefully evaluated when the use of plastic optics is considered. Plastics are well suited for instances where large quantities of finished parts produced by the injection molding process might be involved. A design that incorporates plastics will nearly always be lower in cost, more rugged, and lighter in weight. On the other hand, it will be much more sensitive to temperature changes and will generally exhibit optical quality that is considerably less than that of a similar glass design. Quite often a very successful design will result when the beneficial characteristics of both optical plastic and optical glass are combined. Such a design might incorporate glass elements in areas exposed to weather and frequent cleaning, while plastics would be included to reduce weight and introduce low-cost molded aspheric surfaces.

9.7 Review and Summary

This chapter has presented the reader with an introductory discussion of the special materials that are used in the manufacture of optical components. The section on optical glass has identified a set of common optical glasses that are particularly well suited to the majority of lens designs. The section on low-expansion materials describes the characteristics of several optical glasses that have a very low coefficient of thermal expansion. While some of these materials may be used for lenses and prisms, their primary use will be as a substrate for first-surface mirrors.

Whenever light passes through an air-to-glass interface there will be losses due to surface reflections. The subject of antireflection (AR) coating has been presented, with data detailing the effectiveness of these coatings.

For infrared systems, special and unique optical materials will be required. The most common of these IR materials have been described and their most significant optical and physical characteristics presented. Finally, the subject of optical-grade plastics has been covered. A discussion of the pros and cons of plastics in optical system design is included.

The table of data shown in Fig. 9.11 summarizes the most significant properties of the IR materials and optical plastics that have been discussed in this chapter.

Material	Symbol	Density (g/cm³)	Thermal Exp. (α/°C)	dn/dt (/°C)	Index of Refraction (Wavelength in μm)								
					.586	.485	.656	3.0	4.0	5.0	8.0	10.0	12.0
(IR)													
Germanium	Ge	5.32	6.0×10^{-6}	40.8×10^{-5}	---			4.045	4.0245	4.016	4.0055	4.0032	4.0020
Silicon	Si	2.33	4.2×10^{-6}	3.9×10^{-5}	---			3.432	3.425	3.422	3.4184	3.4179	3.4176
Zinc Sulfide (Multi Spectral)	ZnS	4.09	7.9×10^{-6}	4.1×10^{-5}	2.371	2.439	2.342	2.257	2.252	2.246	2.223	2.200	2.170
Zinc Selenide	ZnSe	5.27	7.6×10^{-6}	6.1×10^{-5}	2.631	---	2.576	2.438	2.433	2.430	2.417	2.407	2.393
(Plastics)													
Methyl Methacrylate	Acrylic	1.19	6.5×10^{-5}	-8.5×10^{-5}	1.491	1.497	1.489	---					
Polystyrene	Styrene	1.06	6.3×10^{-5}	-12.0×10^{-5}	1.590	1.604	1.584	---					

Figure 9.11 Table of significant optical and mechanical properties of IR materials and optical plastics discussed in this chapter.

Chapter 10
Visual Optical System

10.1 Introduction

This chapter will be devoted to a description of the human eye that will provide a basic understanding of how the eye functions and how it will interact with visual optical systems. Also covered will be the topic of optical design and lens design as they apply specifically to systems that are used with the eye as the final detector. Because the human eye is a living and dynamic component, it is important that the optical designer be aware of some details regarding its construction and function. Each individual is unique in terms of his or her specific eye characteristics. As a result, the visual optical system must be designed with all of those potential differences in mind.

10.2 Structure of the Eye

Figure 10.1 shows a cross section view through the typical human eye. This is a simplified version, suitable for the purpose of understanding the basic function and makeup of the eye. The eye is nearly spherical in shape with a diameter of approximately 1 in. Because it is filled with a jelly-like substance, the eye is flexible (similar in feel to a tennis ball). The front of the eye contains the *cornea*, a convex transparent window about 12 mm in diameter that allows light to enter the eye. Because of the cornea's curvature, it is responsible for most of the lens power that is present in the eye. The volume of the eye between the cornea and the *eyelens* is filled with a clear fluid called the *aqueous*. Located just in front of the eyelens is the *iris*, an opaque layer with an opening at its center that will adjust in diameter, depending on the amount of light that is present in the scene being viewed. The diameter of the iris opening will vary over a range of 2 to 7 mm. The normal size, for typical daylight viewing, is about 3 mm.

While the *eyelens* is quite powerful (EFL ≈ 9.6 mm in air), it is considerably less powerful in its actual environment due to the fact that it is immersed in a fluid that has nearly the same index of refraction as the lens itself. The cornea is found to contribute about 60% of the total power of the eye (34 D), while the lens contributes the remaining 40% (23 D). A key function of the eyelens is its ability

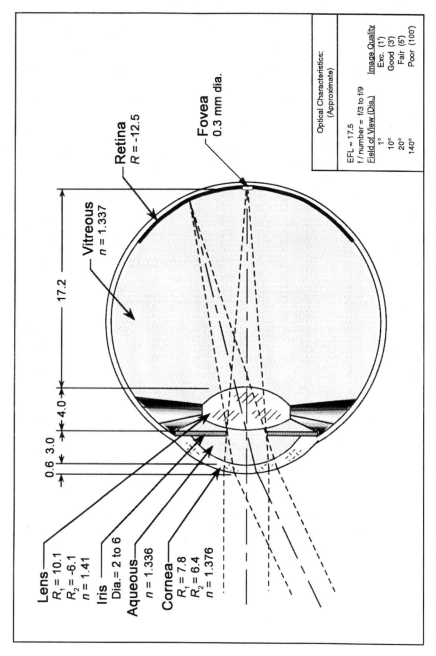

Lens
$R_1 = 10.1$
$R_2 = -6.1$
$n = 1.41$

Iris
Dia.= 2 to 6

Aqueous
$n = 1.336$

Cornea
$R_1 = 7.8$
$R_2 = 6.4$
$n = 1.376$

Retina
$R = -12.5$

Vitreous
$n = 1.337$

Fovea
0.3 mm dia.

17.2

4.0

0.6 3.0

Optical Characteristics:
(Approximate)

EFL = 17.5
f / number = f/3 to f/9
Field of View (Dia.)   Image Quality
       1°   Exc. (1')
      10°   Good (3)
      20°   Fair (5')
     140°   Poor (100')

Figure 10.1 The human eye is a sophisticated optical instrument, quite similar in its function to a camera. The above representation is a simplified version of the typical eye. The actual eye is considerably more complex, and its characteristics will vary slightly from person to person.

to change shape and, as a result, its lens power, thus allowing the average young eye to focus on objects at distances from infinity down to 10 in. This ability of the eye to change focus is referred to as *accommodation*.

The eyelens is followed by a relatively large distance to the focal surface of the eye, which contains the *retina*. The volume between the eyelens and retina is filled with another clear, water-like fluid called the *vitreous*. The retina, like the film in a camera, is the light-sensitive surface on which the image of the eye is formed. The retina is a concave surface with a radius of approximately 13 mm. The *fovea* is a small spot near the center of the retina where the highest level of visual acuity is present. Components of the retina are capable of converting the image information to electrical signals which are then transported to the brain via the optic nerve.

Figure 10.2 shows the approximate field of view for each eye and the combined field of both. While the active area of the retina is large enough to give each eye up to a 140-deg+ field of view, the area over which we can resolve fine detail is limited to just a few degrees. Whenever an object of interest is detected near the outer edges of the field of view, the head and eye will be quickly rotated to bring that object to the center of the eye's field of view for closer scrutiny. Figure 10.3 illustrates the loss of visual acuity as a function of field angle.

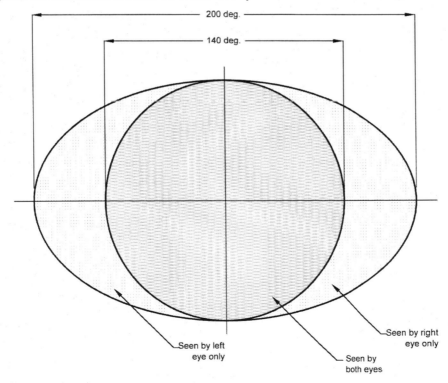

Figure 10.2 Approximate visual field of view. The central 140-deg circle is the binocular portion of the field.

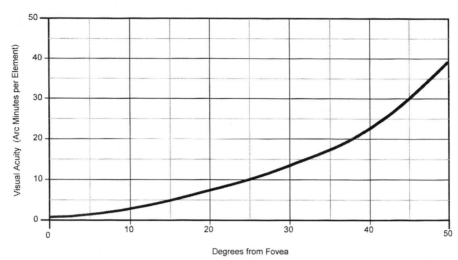

Figure 10.3 Visual acuity (resolution) of the eye is very much dependent on the location of the image on the retina. Maximum acuity is achieved over a very small field, limited by the size of the eye's fovea.

The curvature of the outer surface of the cornea is the key factor responsible for correct vision. When that curvature is too little or too great, the result will be *hyperopia* (farsightedness) or *myopia* (nearsightedness). *Astigmatism* results when the cornea's surface is toroidal rather than spherical. These vision defects can generally be easily and effectively corrected by eyeglasses or contact lenses. Recently, surgical procedures have been developed to modify the shape of the cornea's outside surface to correct these vision defects.

With age, the flexibility of the eye's internal lens lessens and the amount of accommodation that is possible is reduced. Eventually, around age 60, the eye's accomodation is eliminated entirely. A more serious problem, also associated with aging, involves cataracts that cause the eyelens to become cloudy, reducing its transmission and causing scattered light that impairs vision severely. Here again, relatively straightforward surgical procedures have been developed where the lens (with cataract) can be removed and replaced with an implanted artificial lens to restore normal vision.

10.3 Resolution of the Eye

This section will cover, in greater detail, the resolution or visual acuity of the typical eye. The discussion will be limited to the central few degrees of the field of view where the ability to resolve detail is at its maximum.

In the most common test for visual acuity the observer is placed at a known distance (commonly 20 ft) from a test chart containing several rows of block letters, each row of letters progressively smaller in size. A person with normal vision will be able to resolve those letters that subtend a vertical angle of 5 min of arc. At 20 ft, a letter that is 8.9 mm in height will subtend that 5 min angle (see Fig. 10.4).

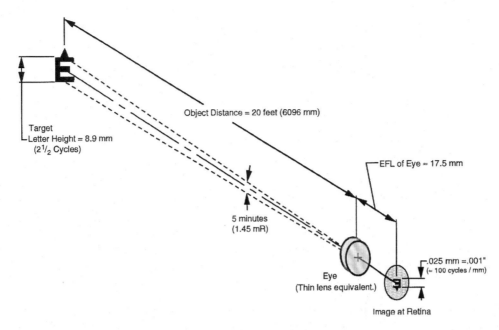

Figure 10.4 Conditions of the standard eye test for visual acuity. When the block letter that is 8.9-mm high is resolved by the eye at a distance of 20 ft, the viewer is said to have normal, or 20/20 vision.

In the field of optical engineering, it is more common to designate resolving power in terms of cycles per millimeter or, for angular resolution, cycles per milliradian (mrad). The block letter of the visual test chart may be thought of as being made up of five horizontal strokes, or 2½ cycles of a repeating pattern of black and white bars of equal thickness. For the example shown in Fig. 10.4, the spatial frequency of the target letter is 2.5 cycles/8.9 mm = 0.28 cycles/mm. The angular frequency of the same letter, at a distance of 20 ft, is 2.5 cycles/1.45 mrad = 1.72 cycles/mrad.

By applying the established principles of thin-lens analysis and the known optical characteristics of the typical eye, we can construct the example illustrated in Fig. 10.4. Here we have a visual resolution (acuity) test chart at a distance of 20 ft and a normal viewer, whose eye is forming an image of that chart on the retina. As we can see from the figure, the image is inverted. Fortunately, the image processing portion of the brain is able to deal with this and we perceive the scene as being erect. The geometry of this example leads to the conclusion that the image of the 8.9-mm-high block letter **E** will be about 0.025 mm, or 0.0001-in. high on the retina. In more familiar optical design terms, this converts to a visual resolution capability of 1.72 cycles/mrad or 98 cycles/mm at the eye's retina.

Figure 10.5 contains an aerial image modulation (AIM) curve for the retina (at the fovea) and MTF curves for the optics of the typical eye (pupil diameter = 3 mm) when it is diffraction limited and, more typically, when there is a small amount of image degradation. The AIM curve indicates the modulation (contrast) that must be present in the image at the fovea in order for the pattern to be resolved. The MTF curve indicates the modulation that will be produced by the optics of the eye for that image. These curves represent typical data and will tend to vary slightly from person to person. From Fig. 10.5 it can be seen that at a frequency of 1.72 cycles/mrad, which we said corresponds to the standard (20/20) visual acuity test target, the fovea requires a modulation of 0.20, while the eye's optics are delivering an image with a modulation of 0.45. The result is that the 20/20 target will be resolved. The next line of letters on the standard visual acuity chart is labeled 20/15 and contains letters that are smaller by a factor of 0.75×, resulting in an angular frequency of 2.3 cycles/mrad (at a distance of 20 ft). At this frequency it can be seen from Fig. 10.5 that the fovea requires a modulation of 0.80 for resolution, while the eye's optics are delivering an image with a modulation of only 0.35. The result is that the 20/15 target will not be resolved. It should be the goal of the optical engineer that the degradation of the eye's MTF due to the introduction of that instrument will be kept to a minimum when designing an optical instrument to be used with the eye.

10.4 Visual Instrument Design Considerations

An example of the procedures used to generate the design of a basic visual instrument will be helpful in illustrating a few of the major considerations involved. First, it is essential that the actual application and requirements of the

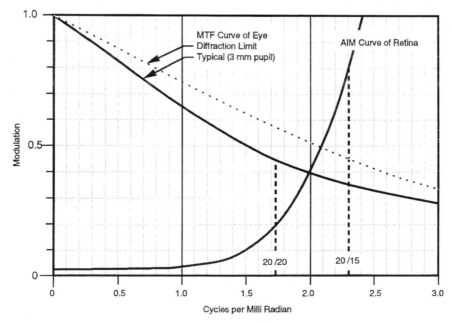

Figure 10.5 The resolution of the eye can be represented in several ways. Shown here is an aerial image modulation (AIM) curve for a typical retina, at the fovea, under normal viewing conditions. When combined with the MTF curve of the average eye's optics, it is possible to estimate the ultimate resolution of the combination (retina plus eye optics).

instrument be clearly defined and understood. For this example we will assume a hypothetical case where it is required that we be able to observe visually and record motor vehicle license plate numbers from a distance of 100 yd. Starting data indicate that the height of such numbers (or letters) is typically 3 in., and that each figure can be assumed to consist of five horizontal strokes, similar to the visual acuity test target described earlier. While it is perfectly logical and correct to describe the target distance in yards and the letter height in inches, our calculations will be simplified if we convert all dimensions to the same units; in this case we will use millimeters. From Fig. 10.6 we can see that the typical target letter will subtend an angle of $76.2/91440 = 0.83$ mrad. It was determined earlier that in order to visually resolve this type of target it must subtend an angle of at least 1.45 mrad. This confirms the need for an optical instrument to magnify the apparent size of the target in order to observe and resolve the required detail at that target. The introduction of a telescope with a magnification of 2× would theoretically produce an image that would be adequate for these purposes. However, if no other obvious negative factors are apparent, it would be prudent to consider a telescope with greater magnification to be certain of achieving the required performance.

In Chapter 8 a 10× telescope was described that would seem appropriate to this application. The basic optical parameters of that telescope (taken from Fig. 8.11) have been added to the layout in Fig. 10.6. It can be seen that two images of the target are formed in this system, one within the telescope, and the final

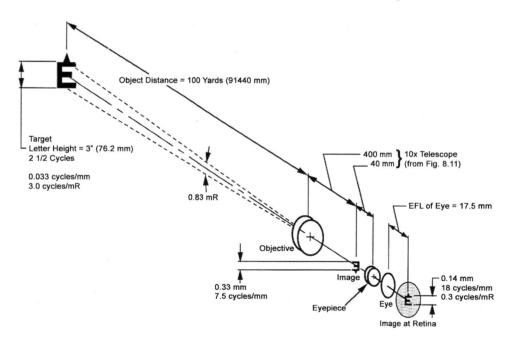

Figure 10.6 A 10× telescope magnifies the angular size of a target such that it is possible to visually resolve 3-in.-high letters at a distance of 100 yd.

image on the eye's retina. It is useful to calculate the image size and spatial frequency at both image planes. Based on earlier analysis, we might conclude that if the spatial frequency of the image at the retina is less than 100 cycles/mm or the angular frequency of the target to the eye is less than 1.72 cycles/mrad, then that target will be resolved. This assumes that the MTF of the eye's optics has not been significantly degraded by the introduction of the optical instrument. In reality, we are introducing an optical assembly between the eye and the target which we must assume contains some amount of residual optical error. Experience indicates that the amount of error present in a precision optical instrument such as this will be in the order of one-half wave of optical path difference (OPD).

Let's digress here for a moment to consider the effect that small amounts of wavefront error will have on the MTF of an optical system. Figure 10.7 contains a normalized MTF curve for a diffraction-limited system. It also shows curves for systems of the same f/# and wavelength when small amounts of wavefront error are present. It is clear from this illustration why a system with ¼λ of error is considered to be essentially diffraction limited. It can also be seen that for an error of ½λ, the MTF cutoff frequency (maximum resolution) will be about 0.5 of the diffraction limit, while a wavefront error of 1λ reduces the maximum resolution to about 0.2, or one-fifth of the diffraction limit.

The corresponding diffraction-limited maximum resolution for the typical eye under standard viewing conditions with a 3-mm pupil is about 300 cycles/mm, which converts to 5.25 cycles/mrad. It is reasonable to assume that the optics of this typical eye might introduce about ¼λ of OPD error. From Fig. 10.7 it can be seen that at 0.5 of the cutoff frequency, a diffraction-limited system

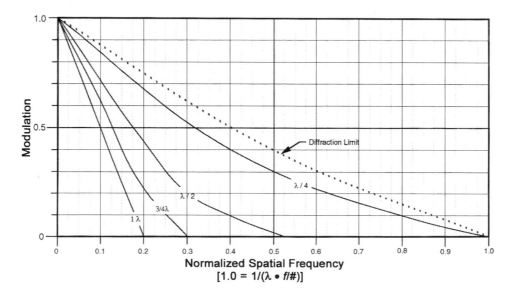

Figure 10.7 Normalized MTF curves for an optical system that shows the reduction in MTF due to small amounts of wavefront error in the system.

will have a modulation of 0.4, while a system with ¼λ error will have a modulation of 0.3. Figure 10.8 shows again the AIM curve for the retina (at the fovea) and the MTF curves for a typical eye, along with one that has acquired an additional ½λ OPD error due to the introduction of the 10× telescope. The critical point to note here is that, while the MTF has clearly been degraded, the telescope has, at the same time, increased the apparent size of the letters being viewed, such that their angular frequency is now 0.3 cycles/mrad. Looking back at Fig. 10.6, we can see that the angular frequency of the target letters to the naked eye is 3.0 cycles/mrad, which is beyond the resolution limit of the retina (≈2.4 cycles/mrad). However, when the 10× telescope is introduced, the angular frequency of the letters is reduced to 0.3 cycles/mrad. At this frequency, from the curve shown in Fig. 10.8, it can be seen that the retina requires a modulation of about 0.03, while the eye plus telescope combination (the ¾λ curve) deliver a modulation of 0.78. As we anticipated, the letters viewed under these conditions will be easily resolved.

10.5 Visual Instrument Focus

Focus adjustment in a visual instrument is required in order to accomplish two purposes. The first of these is to vary the object distance at which objects are in

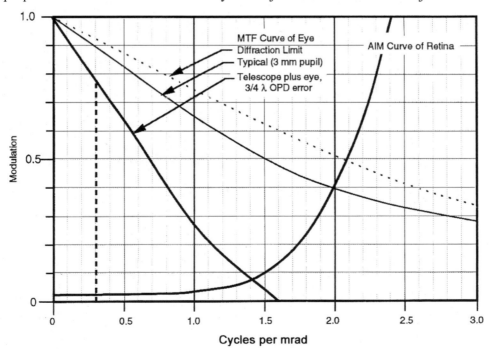

Figure 10.8 The resolution of the eye will be degraded by the addition of an optical system, in this case a 10× telescope (see Fig. 10.6). While the visual MTF is degraded as shown by the ¾λ curve, the angular frequency of the target has been reduced from 3.0 to 0.3 cycles/mrad. As the curves show, at that frequency (0.3) the required image modulation at the retina is just 0.03, while the optics of the telescope plus the eye will deliver a modulation of 0.78, making the target easily resolvable.

focus, while the second is to compensate for the viewer's individual eyesight condition in terms of near or farsightedness. For instance, in the case of the 10× telescope in the example just described, the focus adjustment might be established as follows. First, the application specified an object distance of 100 yd. We might reasonably assume that this telescope should be capable of focus over a range from infinity down to 25 yd. From earlier thin-lens exercises we will recall the Newtonian equation, which can be used to determine the focus shift relative to the object distance:

$$x \cdot x' = f^2,$$

where x is the object distance, x' is the focus shift, and f is the objective lens focal length.

The focal length of the telescope's objective lens in this case is 400 mm. Using the thin-lens formula, an object distance of infinity will result in a focus shift of zero. For the specified viewing distance of 100 yd (91,440 mm), the focus shift will be 1.75 mm and, for the minimum object distance of 25 yd (22,860 mm), the focus shift will be 7.0 mm. Figure 10.9 (top) shows the image plane location for each of these cases. If the telescope were designed such that the eyepiece could focus on the infinity image plane and then be adjusted (away from the objective) over a distance of 7.0 mm, this would accomplish the goal of being able to focus the telescope on objects from infinity down to 25 yd.

The eyepiece adjustment to compensate for viewer differences is generally described in terms of diopters (D). While specifications will vary, a typical requirement will call for an instrument that is capable of being adjusted over a range of from –4 to +2 D. This is the range that will be used for this example. The amount of eyepiece travel required to produce a 1.0-D focus shift is equal to $f^2/1000$, where f is the focal length of the eyepiece in millimeters. For the eyepiece in this example the focal length is 40 mm, resulting in a focus travel requirement of $40^2/1000 = 1.6 \, \text{mm/D}$. For a range of –4 to +2 D, the eyepiece will have to be adjusted from –6.4 mm to +3.2 mm from its nominal, zero-diopter setting.

Referring again to Fig. 10.9, it can be seen that the eyepiece focal point will have to vary from a point 6.4 mm inside the infinity focus to a point 3.2 mm beyond the 25-yd focus, for a total travel of 16.6 mm. This amount of eyepiece travel will allow for the change in object distance as well as the viewer accommodation required for the majority of individuals. This is a specification (focus travel) that should always be set on the generous side to ensure that the basic requirement will be met with some degree of over travel. In this case, an actual overall total eyepiece travel of 20 mm (+12/–8) would not be unreasonable.

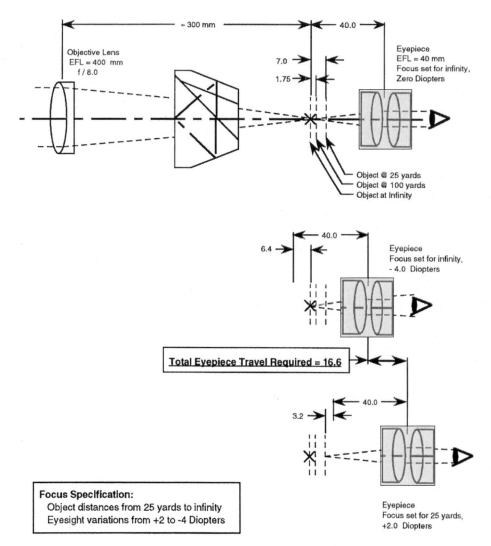

Figure 10.9 Instrument focus must accommodate a range of object distances and also eyesight variations that may exist among those individuals using the instrument.

10.6 Visual Instrument Detailed Lens Design

Having covered the basic structure, function, and performance of the eye, it will be useful at this point to consider the details of the lens design procedures used to create an optical instrument that will be used with the eye as the final detector. Referring back to the 10× telescope described in Chapter 8 (Fig. 8.11), this section will present some of the detailed considerations that might be involved in generating an actual final lens design for this telescope.

There will be three major components that the optical engineer must consider in order to produce a complete optical design for this telescope: the objective lens, the roof Pechan prism, and the eyepiece. Starting with the objective lens, the basic specification from Chapter 8 calls for a lens diameter of 50 mm, a focal length of 400 mm, and a 5-deg field of view. For a 10× telescope, this will result in a 50-deg apparent field of view at the eye. While this wide field of view is good for certain applications, such as sporting events and bird watching, it is a wider field than is required for a typical surveillance instrument, such as the telescope we need to read license plates at 100 yd. The objective lens design will be simplified by assuming that the field of view for this example will be 4 deg (40 deg at the eye). The starting lens form might be selected from any one of numerous sources. In Chapter 7, a lens of these basic specifications was presented (see Fig. 7.5). We will use that lens as a starting point for this design. The correction of this lens will need to be modified to account for the roof Pechan prism that must be inserted into its back-focus region in order to produce a telescope with correct image orientation. If we assume a roof Pechan prism with a 40-mm aperture, the glass path through the prism will be 6 × 40 = 240-mm long.

The objective lens design will be reoptimized to include this 240-mm glass path using the lens curvatures and focus as variables. The primary design constraints that must be controlled during the optimization procedure are a focal length of 400 mm and the correction (or balance) of spherical aberration and chromatic aberration to produce essentially diffraction-limited performance on axis. The off-axis aberrations will be monitored and kept under control, but they will definitely assume a secondary role during the design optimization process.

As the starting doublet design is shown in Fig. 7.5, it has an on-axis blur circle radius of 0.003 mm, indicative of essentially diffraction-limited performance (the Airy disk radius is 0.0055 mm). Adding the 240-mm thickness of BK7 to simulate the prism does disturb the state of correction, particularly the on-axis color, resulting in a blur circle with a radius of 0.012 mm, about twice the diffraction limit and an unacceptable value for this application. The doublet design is reoptimized to produce a lens-prism combination that is free of spherical aberration, primary axial color, and coma. During optimization it was noted that the air space was contributing little to performance, so the design was changed to a cemented form. Aberration curves for the resulting final design are shown in Fig. 10.10. It is important to keep in mind that the image produced by this lens will be viewed by the eye using a 40-mm-focal-length (≈6.3×) eyepiece. Examining the aberration curves, we see that the on-axis (FOB 0) curves are close to zero, essentially diffraction limited. Looking at the off-axis curves, we see that they are tilting, indicating a focus error (field curvature and astigmatism). We also see that the off-axis curves are separating into three parallel curves, indicating off-axis or lateral color. It is at this point that the fact that we are designing a visual system becomes significant. The focus problem, for example, will not really be a problem, as it can be easily accommodated (refocused) by the eye. This eye focus is simulated by designing an image surface that is curved rather than flat. The astigmatism, or field curves, in Fig. 10.10 show us that the maximum amount of focus error due to astigmatism (separation of the *T* and *S*

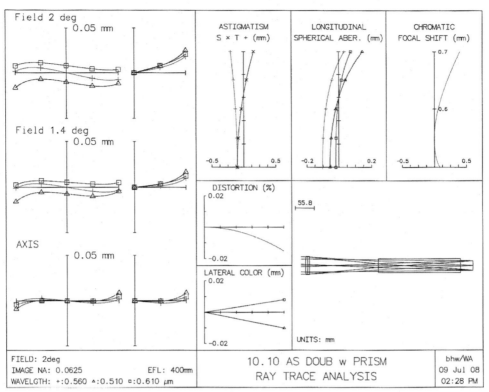

Figure 10.10 Aberration curves plus a lens drawing for a 400-mm, *f*/8 achromatic cemented doublet, designed to operate with a 40-mm-aperture roof Pechan prism in its back-focus region to provide an erect image.

curves) is about +/–0.20 mm. From our earlier calculation of 1.6 mm per diopter for the 40-mm eyepiece being used, these numbers represent focus errors of just +/–0.12 D. This amount of residual astigmatism will not be a problem for a normal, healthy eye. The curves for lateral color show us that the maximum amount present is about 0.018 mm. When viewed through the 40-mm eyepiece, this represents an angular blur of $0.018/40 = 0.0005$, or less than 2 min of arc to the eye. This is an acceptable amount of lateral color that will not be noticeable when viewing most objects. The final curve in the ray-trace analysis (Fig. 10.10) is the distortion curve. At less than 0.02%, distortion is obviously not a problem. The end product of this design exercise is a cemented doublet achromatic objective lens with a focal length of 400 mm, a speed of *f*/8, capable of imaging a 4-deg circular field of view, with an image quality that is quite acceptable for the 10× visual telescope application being considered. This lens will perform as predicted only when it is used in conjunction with a prism made from BK7 optical glass with a total internal path length of about 240 mm.

The prism design will be based on the standard roof Pechan prism configuration, as shown in Fig. 10.11. In addition to detailed manufacturing tolerances, the prism drawing should carefully spell out the overall optical quality required by specifying the maximum allowed wavefront error for a transmitted collimated light beam equal in diameter to the prism aperture. In this case a specification of ¼λ single-pass maximum wavefront error for a 40-mm-diameter beam will assure performance that is consistent with our telescope requirements.

The final optical component of the telescope is the 40-mm focal length eyepiece. There are many sources for eyepiece designs, including *MIL Handbook 141*, patent literature, and numerous textbooks. An eyepiece design that will work well in this application can be found in my book *Optical Design for Visual Systems* on page 57.[1] That design is a modified symmetrical form with a focal length of 28 mm. This lens data can be taken into the computer and scaled to the required 40-mm focal length using a subroutine that is a part of the OSLO lens design package. The resulting lens data for the eyepiece can then be merged with the objective lens design that was just generated to form the complete 10× telescope. Fine tuning of the combined lens data includes adjusting the prism-to-eyepiece distance for best focus and sizing the eyepiece element diameters to ensure that the full off-axis light bundles will be transmitted. The telescope analysis allows us to locate the system exit pupil, assuming the entrance pupil to be in coincidence with the first surface of the objective lens. The resulting eye relief (eyepiece to exit pupil distance) is found to be 30 mm, a comfortable dimension for viewers with or without eyeglasses.

The mechanical design of this telescope will include a 28-mm-diameter field stop at the internal image plane, along with several glare stops, or baffles, to eliminate stray light that might enter the objective from outside the 4-deg field of view and find its way to the eye. A convenient location for these baffles will be at the entrance and exit faces of the roof Pechan prism. This choice of location will make it possible for the baffles to be made an integral part of the prism's mechanical mount. Figure 10.11 shows the final telescope design configuration. Computer analysis of this design shows it to be diffraction limited on axis, with about 2 D of astigmatism and 4 min of lateral color at the maximum field angle (these are combined objective plus eyepiece aberrations). The eyepiece also introduces about 5% distortion. All these values are indicative of a visual system

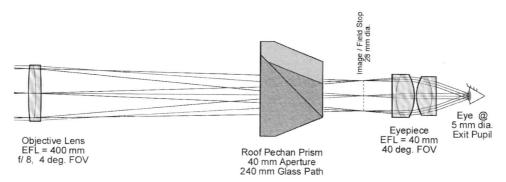

Figure 10.11 The telescope design shown has been evaluated and found to be compatible with the function and performance of the eye.

of high quality and very good overall optical performance. Figure 10.12 contains tabulated data on the final telescope design, as generated by the OSLO optical design software package.

10.7 Review and Summary

The material contained in this chapter has been presented to introduce the reader to the basic characteristics of the human eye as they might affect the optical design of instruments intended for visual applications. The general structure of the eye has been described and an optical model presented that illustrates the

```
*PARAXIAL CONSTANTS
    Angular magnification:    -10.00000    Lagrange invariant:      -0.87302
    Eye relief:                30.0000     Petzval radius:         -54.76645

*LENS DATA
10.11 Telescope
  SRF      RADIUS       THICKNESS    APERTURE RADIUS        GLASS  SPE  NOTE

  OBJ        --         1.0000e+20   3.4921e+18            AIR

  AST     262.50000 V    7.00000     26.00000 A            BK7  C  Objective
   2     -154.60000 V    5.00000     26.00000              F2   C
   3     -514.00000 V  200.00000     26.00000              AIR

   4        --         240.00000     20.00000 K            BK7  C  Prism
   5        --          36.85000     20.00000              AIR

   6        --          26.36000     14.00000              AIR   *  Image

   7        --           4.00000     20.00000              SF5  C
   8      42.50000      17.00000     20.00000              BAK1 C  Eyepiece
   9     -42.50000       1.40000     20.00000              AIR

  10      37.00000      14.25000     18.00000              BAK1 C
  11     -37.07172       4.00000     18.00000              SF5  C
  12        --          30.00000     18.00000              AIR

  13        --            --          2.50000              AIR   *  Exit Pupil

  IMS        --     V     --     V                               *
```

```
*WAVELENGTHS
 CURRENT  WV1/WW1    WV2/WW2    WV3/WW3
    1     0.56000    0.51000    0.61000
          1.00000    1.00000    1.00000
```

```
*REFRACTIVE INDICES
 SRF   GLASS        RN1        RN2        RN3        VNBR        TCE
  0    AIR        1.00000    1.00000    1.00000      --          --
  1    BK7        1.51803    1.52077    1.51591    106.59032   71.00000
  2    F2         1.62262    1.62851    1.61821     60.43499   82.00000
  3    AIR        1.00000    1.00000    1.00000      --        236.00000
  4    BK7        1.51803    1.52077    1.51591    106.59032   71.00000
  5    AIR        1.00000    1.00000    1.00000      --        236.00000
  6    AIR        1.00000    1.00000    1.00000      --        236.00000
  7    SF5        1.67584    1.68309    1.67047     53.55423   82.00000
  8    BAK1       1.57402    1.57742    1.57141     95.50385   76.00000
  9    AIR        1.00000    1.00000    1.00000      --        236.00000
 10    BAK1       1.57402    1.57742    1.57141     95.50385   76.00000
 11    SF5        1.67584    1.68309    1.67047     53.55423   82.00000
 12    AIR        1.00000    1.00000    1.00000      --        236.00000
 13    AIR        1.00000    1.00000    1.00000      --        236.00000
```

Figure 10.12 Tabulated computer output from OSLO describing the 10× telescope shown in Fig. 10.11, including paraxial constants and lens prescription data.

form and function of the individual components of the eye. The optical quality of the eye's optics and the retina were introduced so that the approximate resolution capability of the typical eye could be understood. The goal in describing the eye and its performance has been to assist the optical engineer in creating designs that are compatible with the eye. All data presented dealing with the eye has been classified as typical, taken from several sources and simplified to facilitate this analysis. Variations between individuals will be significant, making a more precise description inappropriate.

The final sections of the chapter have been devoted to a general discussion of the optical design of visual instruments, followed by a detailed lens design procedure as it might be used to generate the optical design of a 10× visual telescope. Analysis of that design has been reviewed, with particular attention to the unique aspects of systems intended for visual use.

Reference

1. B. H. Walker, *Optical Design for Visual Systems*, SPIE Press, Bellingham, Washington (2000).

Chapter 11
Lens Design and Image Evaluation

11.1 Introduction

This chapter will introduce the reader to the process of lens design and image evaluation. While not everyone working in the field of optical engineering will be called on to execute detailed lens designs, a basic understanding of the procedures involved and their limitations will be helpful to all. In the area of image evaluation, some degree of familiarity with the various methods employed will allow the optical engineer to more effectively specify required performance, and later, to determine whether those levels of performance have been met by the designer. This chapter is particularly relevant when we consider the state of the computer and optical software field today. The days when access to these tools of lens design was limited to a very carefully chosen few are gone forever. In Chapter 5 we introduced the OSLO and OSLO-EDU optical design software packages, pointing out their ready availability and ease of use in generating and evaluating certain (relatively simple) optical systems. This chapter will expand into the use of a more sophisticated software package (OSLO *Premium Edition*, Rev. 6.1), outlining its use in generating detailed lens design solutions for a variety of specific problems. The field of optical design software is broad and very dynamic. There are a number of very fine software packages to choose from and they are all (including OSLO) constantly updated to reflect the state of the art.

11.2 Lens Design Process

While there should always be flexibility in the process of generating a final lens design, there are a number of steps that are usually involved. These steps are listed below. In the lens design examples that follow, these steps will be followed and described in more detail.

The major steps in the lens design process are as follows.

(1) Define the problem (in optical terms) and establish requirements.
(2) Select a lens design to be used as a starting point.
(3) Modify the starting design, such that it meets all basic requirements.
(4) Optimize the modified starting design such that it conforms to general requirements (EFL, etc.) while providing the best possible image quality.
(5) Evaluate performance and compare with requirements; if good enough, go on. If not, go back to Step 2.
(6) Document the final design (with tolerances).

11.3 10× Telescope Design

11.3.1 Defining the problem

While it may sound trite, the definition phase of the lens design process is critical. Often the ultimate user of the lens system (the customer) can define the problem to be solved only in general terms, terms that are not always consistent with those found in a true optical specification. The optical engineer must then work with the customer to generate a specification that all will be comfortable with. For example, in the last chapter we discussed a situation where the customer needed to be able to read the numbers on a license plate at a distance of 100 yd. The problem was that the customer could not resolve those numbers with the naked eye. The optical engineer's task in this case would be to come up with the concept of the 10× telescope (see Sec. 10.6), and then to demonstrate to the customer how the use of this telescope would solve the basic problem. Having agreed on an approach, it will then be possible to generate a lens design specification that will describe in detail the optical performance requirements of that telescope. The lens designer would then be tasked with generating the final lens design, including a tolerance analysis, optical schematic, and detail drawings of all required optical components, suitable for their manufacture.

11.3.2 Selecting a starting point

Once the lens designer has a complete description of the optical system requirements, the next step is to select a basic optical design configuration that will serve as a good and logical starting point for the lens design process. Past experience is perhaps the lens designer's most valuable asset at this stage of the lens design process. In addition, many of today's reference textbooks and optical design software packages will contain a library of basic lens designs. The combination of experience and access to a comprehensive lens design library will usually allow the designer to settle on a starting point quite quickly.

In the example of the 10× telescope, the lens design process can be broken down into two separate steps: the objective lens design (with prism), followed by the selection of an eyepiece design. In this section we will describe in more detail the procedures that the lens designer might have followed in generating the final telescope design. First, it was established that the objective lens should have a 400-mm focal length, a speed of $f/8$, and image a 4-deg field of view. An achromatic, air-spaced doublet lens, described in an earlier chapter of this book, was found to meet these basic requirements and was chosen as a starting point.

11.3.3 Modifying the starting point

The starting point design was examined to determine any basic changes that might be made to make it more suitable to the requirements of this particular design. In this case two changes were introduced: the lens was changed to a cemented form to improve producibility, and a block of BK7 optical glass was added to the back focal region of the lens to simulate the required roof Pechan prism.

11.3.4 Optimizing the modified starting design

At this stage of the design the lens designer sets the lens curvatures and the focus position as variables and uses the optimization routine in OSLO to correct the basic aberrations of the objective lens. Figure 11.1 illustrates the start and finish points of the objective lens optimization process.

The second step in the telescope design involves the eyepiece. It was established earlier that an eyepiece would be required that was capable of covering a 40-deg field of view, with a focal length of 40 mm and a speed of $f/8$.

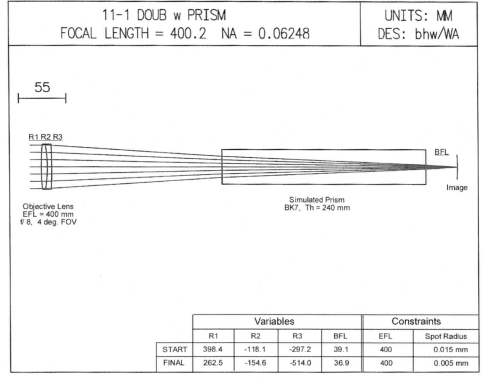

	Variables				Constraints	
	R1	R2	R3	BFL	EFL	Spot Radius
START	398.4	-118.1	-297.2	39.1	400	0.015 mm
FINAL	262.5	-154.6	-514.0	36.9	400	0.005 mm

Figure 11.1 Objective lens for a 10× telescope with the basic variables and constraints that were used during optimization.

A design was taken from a reference text and scaled to the required focal length. Its performance was found to be quite acceptable when combined with the objective lens and prism.

This is a classic example of the use of an existing lens design when possible. In this case, the objective lens design called for the presence of a large prism in its back focal space, making the use of an existing doublet design impossible. In the case of the eyepiece, the specifications and the method in which it was to be used were quite conventional, making it possible to take an existing eyepiece design and simply scale its dimensions to the required focal length and field angle.

11.3.5 Evaluating performance

With the new objective lens and eyepiece combined to make a 10× telescope, the system was analyzed for image quality. It was established that the primary goal of this design was to provide a telescope that would introduce no more than ½λ of error to the on-axis wavefront. One of the easiest and most effective methods of evaluating whether this goal has been met is to execute a diffraction MTF analysis which includes spherical aberration, chromatic (color), and diffraction effects. This was done for this telescope design and the resulting MTF curve is shown in Fig. 11.2. The near coincidence of the actual MTF curve to the

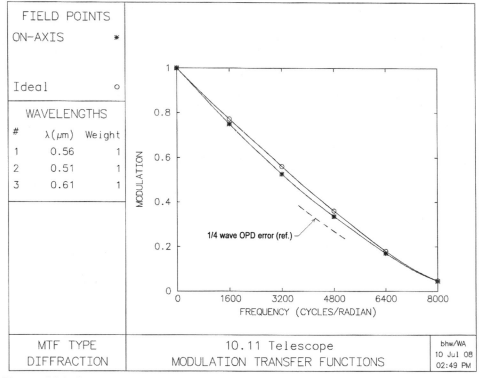

Figure 11.2 On-axis, polychromatic, diffraction MTF curve for the 10× telescope final design. This figure illustrates that residual aberrations are much less than ¼λ.

ideal (diffraction limit) curve shows that the effective wavefront error through the telescope is considerably less than $\frac{1}{4}\lambda$ (est. = 0.07λ). This leads to the conclusion that the majority of the allowable $\frac{1}{2}\lambda$ telescope tolerance can be assigned to the various manufacturing aspects of the design, most particularly the objective lens and the roof Pechan prism.

11.3.6 Documenting the final design

Once a lens design has been generated and found to meet all system requirements, it is the lens designer's final responsibility to document that design thoroughly for purposes of manufacture and for future reference. For most new optical designs the following documentation should be provided:

(1)   tabulated lens data (prescription),
(2)   optical schematic,
(3)   tabulated tolerance data, and
(4)   detail drawings of all optical components.

The tabulated lens data records the nominal design information. Ideally, this table should be generated by the software package that was used to create the design. Prior to generating the tabulated lens data, the designer should examine the final design and round off the values given for lens radii and thickness to a level that is consistent with the manufacturing tolerances. The table of optical data for the 10× telescope, as produced by the OSLO lens design program, is shown in Fig. 11.3.

The optical schematic should show the optical components of the system and establish the relationship (usually spacing) between them, with tolerances where appropriate. Any aperture, field, or glare stops that are required should be identified, located, and sized on the optical schematic. Finally, the optical schematic should contain a table of all important optical system characteristics, such as magnification, field of view, and light transmission. The optical schematic should be a comprehensive reference document, containing any and all optical data that will be frequently referenced during the existence of the optical instrument that has been created. Figure 11.4 is an optical schematic of the 10× telescope that illustrates the typical design data and other information that might be included.

11.4 Precision Collimator Lens Design

11.4.1 Defining the problem

This example will deal with the design of an objective lens for a precision collimator that is to be used as part of a spectrometer system. It is the function of this collimator to collect radiation from a point source and project that energy in the form of a collimated beam. The basic parameters established for this lens are

```
*PARAXIAL CONSTANTS
    Angular magnification:     -10.00000    Lagrange invariant:      -0.87302
    Eye relief:                 30.0000     Petzval radius:         -54.76645

*LENS DATA
10.11 Telescope
 SRF       RADIUS        THICKNESS    APERTURE RADIUS     GLASS  SPE   NOTE

 OBJ         --          1.0000e+20    3.4921e+18          AIR

 AST     262.50000 V      7.00000      26.00000 A         BK7 C  Objective
  2     -154.60000 V      5.00000      26.00000            F2 C
  3     -514.00000 V    200.00000      26.00000           ATR

  4         --          240.00000      20.00000 K         BK7 C  Prism
  5         --           36.85000      20.00000           AIR

  6         --           26.36000      14.00000           AIR   * Image

  7         --            4.00000      20.00000           SF5 C
  8      42.50000        17.00000      20.00000           BAK1 C  Eyepiece
  9     -42.50000         1.40000      20.00000           AIR

 10      37.00000        14.25000      18.00000           BAK1 C
 11     -37.07172         4.00000      18.00000           SF5 C
 12         --           30.00000      18.00000           AIR

 13         --              --          2.50000           AIR   * Exit Pupil

 IMS        --    V         --    V                              *

*WAVELENGTHS
CURRENT   WV1/WW1       WV2/WW2      WV3/WW3
   1      0.56000       0.51000      0.61000
          1.00000       1.00000      1.00000

*REFRACTIVE INDICES
 SRF      GLASS         RN1           RN2          RN3        VNBR         TCE
  0       AIR         1.00000       1.00000      1.00000      --           --
  1       BK7         1.51803       1.52077      1.51591    106.59032    71.00000
  2       F2          1.62262       1.62851      1.61821     60.43499    82.00000
  3       AIR         1.00000       1.00000      1.00000      --        236.00000
  4       BK7         1.51803       1.52077      1.51591    106.59032    71.00000
  5       AIR         1.00000       1.00000      1.00000      --        236.00000
  6       AIR         1.00000       1.00000      1.00000      --        236.00000
  7       SF5         1.67584       1.68309      1.67047     53.55423    82.00000
  8       BAK1        1.57402       1.57742      1.57141     95.50385    76.00000
  9       AIR         1.00000       1.00000      1.00000      --        236.00000
 10       BAK1        1.57402       1.57742      1.57141     95.50385    76.00000
 11       SF5         1.67584       1.68309      1.67047     53.55423    82.00000
 12       AIR         1.00000       1.00000      1.00000      --        236.00000
 13       AIR         1.00000       1.00000      1.00000      --        236.00000
```

Figure 11.3 Tabulated lens data for the 10× telescope, including all nominal information required for generating manufacture and assembly drawings.

a focal length of 300 mm, a speed of *f*/6 (or faster), and a spectral bandwidth of 0.6 to 1.1 μm. Because it is being used to create a well-corrected collimated beam, only the on-axis performance of this lens needs to be considered. The lens specification further states that overall image quality equivalent to less than ½λ wavefront error is required. The unique parameter of this lens, dictating the need for a custom lens design, is clearly the relatively broad near-infrared spectral bandwidth.

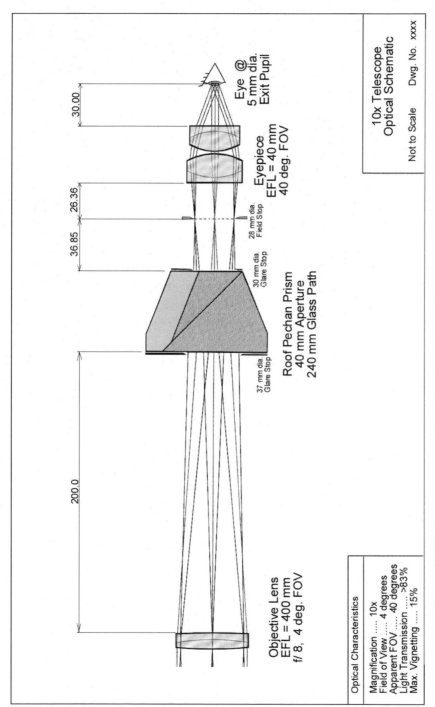

Figure 11.4 Optical schematic of the 10× telescope that shows typical general design information.

11.4.2 Selecting a starting point

At this stage of the design process it is wise for the lens designer to consider the lens requirements in general and to prioritize them in terms of how each might impact the basic lens design. In this case the lens focal length and speed are not unusual and the field of view to be covered is essentially zero. It is probably safe to say that if all the lens requirements were this benign, a stock lens would have been selected and the lens designer's services would not be required. The spectral bandwidth, coupled with the need for near-diffraction-limited image quality, are the factors that will make this a challenging lens design problem.

For any lens or optical system, the spectral bandwidth is determined by the source, the detector, and the transmittance of all that lies between the two. The systems engineer, responsible for the overall design of the spectrometer, might typically provide the lens designer with a spectral response curve similar to that shown in Fig. 11.5. The lens design must be optimized to transmit and image this spectral bandwidth. For purposes of design and analysis the lens designer will need to establish design wavelengths and weights. To do this the designer first divides the spectral band into three bandwidth segments, each covering one third of the total spectral band. Next, a design wavelength and spectral weight is assigned to each segment. As Fig. 11.5 shows, the wavelengths selected for this design will be 0.85, 0.70, and 1.0 μm, with respective weights of 1.0, 0.77, and 0.80. The spectral weight is based essentially on the area under the spectral response curve for each segment.

Because of the relatively wide spectral bandwidth, it was concluded that either an all-reflecting lens or an apochromatic (corrected for three wavelengths) lens design would be required. Problems with physical layout and potential for alignment difficulties led to early elimination of the reflecting lens approach from consideration. The need for an apochromatic design was confirmed by calculations based on the graph that was presented earlier in Fig. 6.9. Since the lens speed of *f*/6 given in the specification is presented as a minimum, we will enter this design process with a goal of *f*/5 for the lens speed for the 300-mm EFL lens (diameter of 60 mm). From Fig. 11.5 we see that the lens will operate at a central wavelength of 0.85 μm, with a spectral bandwidth range of +/–150 nm. Having established these values, we can use Fig. 6.9 to estimate the blur-spot radius due to secondary color in an achromatic lens form at approximately 0.015 mm.

The diffraction-limited blur circle radius (the Airy disk) for an *f*/5 lens at $\lambda = 0.85$ μm will be

$$R = 1.22 \times \lambda \times f/\# = 1.22 \times 0.00052 \times 5 = 0.005 \text{ mm}.$$

A secondary color blur spot that is three times larger than the Airy disk tells us that the achromatic lens form will not perform adequately and that an apochromatic design will be required.

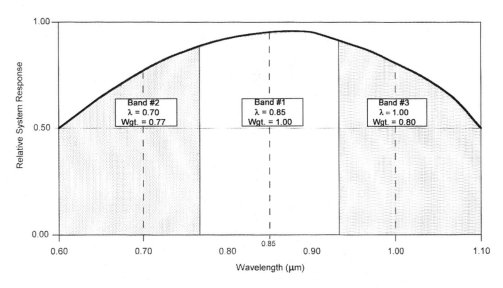

Figure 11.5 System spectral response curve, based on source output, detector sensitivity, and overall system transmittance. Three design wavelengths and weights (based on area) are chosen as shown.

Like most refracting telescope designs, collimator objectives frequently take the form of doublet lenses. In this case, the need for a relatively fast lens suggests the need for an air-spaced doublet to permit adequate correction of spherical aberration. In Chapter 5 (Fig. 5.15) we described an interesting new design for an air-spaced apochromatic diffraction-limited doublet lens. This lens has been selected as the starting point for this design.

11.4.3 Modifying the starting point

The starting lens design will have to be modified in several key ways if it is to be used for our collimator objective. First, the wavelengths for the design must be changed from the original visual band, to the values of 0.85, 0.70, and 1.0 μm, with the spectral weights shown in Fig. 11.5. Next, the entrance pupil radius for the lens must be changed to 30 mm. Finally, the last surface of the lens is assigned paraxial solves for its curvature and for the distance to the image plane. The curvature solve is set so that the angle of the emerging paraxial marginal ray is –0.10, and the final thickness solve is set so that the height of that ray is zero on the next surface (the image). These changes result in a lens that meets the basic system requirements of 300-mm EFL and a speed of *f*/5.0. The only remaining problem is image quality.

A quick look at the modified starting design indicates the blur spot radius for this lens form to be 0.17 mm, more than 30 times greater than the size of the Airy disk (our design goal for this lens). The optimization stage will tell us whether it is possible to manipulate this basic lens form such that it will deliver the required final image quality.

11.4.4 Optimizing the modified starting design

The optimization process involves the manipulation of available variables while monitoring the performance characteristics of the lens, particularly its image quality. For the apochromatic doublet, the four-lens surface curvatures and the focus position were used as variables. Constraints assigned during optimization held the focal length to 300 mm and the lens speed to $f/5$, while minimizing the on-axis chromatic spot size. After several cycles of the lens design process, an optimized configuration for this lens was generated. Figure 11.6 shows the start and finish configurations for this lens. The optimization of this doublet lens was accomplished by using the variable radii to alter the power of the individual elements while maintaining the sum of the powers constant (EFL = 300 mm). At the same time, the shapes of the elements were varied, along with the final image plane location (focus). A final step in the optimization process was to reduce the final lens speed to $f/5.4$. This speed was still well within the specified requirement of $f/6$, and it did positively impact the final spot size and overall lens cost. The spot size that has been achieved (0.006-mm radius) indicates that an acceptable design has been generated. More thorough image evaluation will be done in the next stage of the design process to confirm this conclusion.

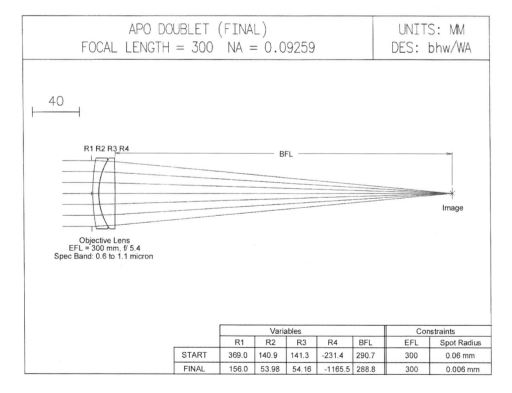

Figure 11.6 The objective lens for a near-IR collimator is shown with the basic variables and constraints that were used during optimization.

11.4.5 Evaluating performance

The performance evaluation of this lens will cover two very basic and equally important areas. First, the basic lens parameters, including manufacturing tolerances and other aspects of producibility, must be considered. Second is the matter of image quality. Any design that measures up in one of these areas and falls short in the other must be considered a failed lens design effort. A lens that is easy and inexpensive to manufacture, yet has poor image quality, represents a poor design. Likewise, a lens with theoretically superb image quality that cannot be produced within the established budget and schedule constraints cannot be considered acceptable.

The apochromatic doublet design presents several manufacturing challenges that cannot be ignored. The most obvious of these is the choice of optical glass type for the positive element. In order to achieve the required apochromatic state of correction in a doublet lens form it is essential that at least one unusual glass type be used. In this case the positive (crown) element is made from the OHARA glass type designated S-FPL51 (497816). This glass is relatively expensive, not readily available, and quite difficult to work with in the optical shop because of an unusually high thermal expansion coefficient. On the other hand, the design requires just a single element of this material, weighing just a few ounces (optical glass is sold by weight), and the introduction of this glass into the design has been clearly crucial to its success. The thermal expansion might be a problem were the instrument to be used under extreme environmental conditions, such as those that might be found in military or space applications. Fortunately, this is not the case for the spectrometer in question, which is intended primarily for use in the laboratory. The mechanical designer responsible for the lens cell design work must take all glass characteristics into account while doing that design work. All in all, the tradeoffs in favor of using this particular glass type, in this application, would seem to prevail.

Optical performance, in terms of spot size, had been monitored carefully throughout the design-optimization phase. Now is the time for a more rigorous analysis of image quality. First, the aberration curves will tell the lens designer a great deal about residual aberrations in the design and how the various aberrations have been balanced to produce the reported spot size. Figure 11.7 (upper left) shows the on-axis ray-intercept curves for this lens at the three principal wavelengths. These curves allow the designer to estimate a blur-spot radius of about 0.004 μm. In the upper right, the corresponding wavefront (OPD) curves are shown. Here, OPD errors of about $\frac{1}{4}\lambda$ indicate the required near-diffraction-limited performance. The lower left quadrant contains the spot diagram and the radial energy distribution (RED) curve is shown in the lower right. Recalling that the Airy disk radius for this lens was 0.005 mm, this spot diagram analysis indicates essentially diffraction-limited performance for this lens.

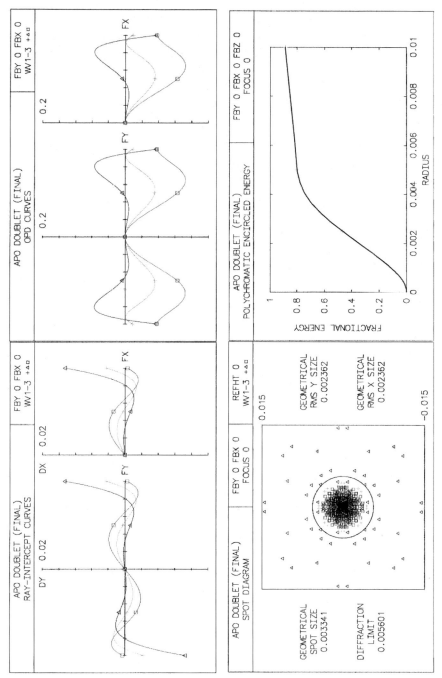

Figure 11.7 Image analysis including ray intercepts, OPD curves, spot diagram, and RED for a 300-mm EFL *f*/5.4 apochromatic doublet, designed to cover a broad spectral band in the near-IR region.

This level of performance is ultimately and conclusively confirmed by the polychromatic diffraction MTF data that is shown in Fig. 11.8. This data indicates about $\frac{1}{20}\lambda$ of residual OPD error. Having achieved all the established performance goals, our next step in the design process is to document the final design for manufacture and for future reference.

11.4.6 Documenting the final design

The first step in the design documentation process is to generate a table of lens data. It will assure accuracy if this is done using the lens design software at the same time that the image quality evaluation has been done. Obvious, but very important, is the fact that this approach guarantees that a lens produced per the tabulated data will yield image quality that is consistent with results of the just-completed analysis. The apochromatic doublet is completely described by the tabulated data presented in Fig. 11.9. This information represents typical lens data as it is output from the OSLO lens design program. From this table it should be possible to extract all nominal information required to manufacture the lens elements and to fabricate the lens assembly.

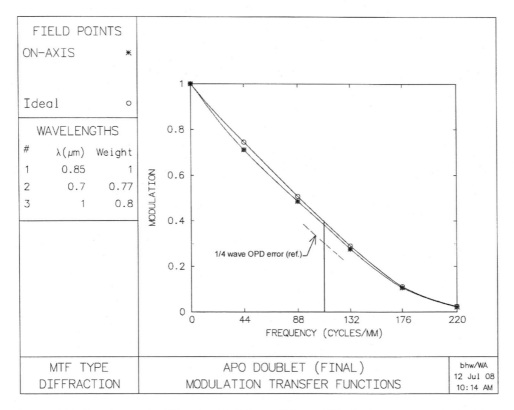

Figure 11.8 Polychromatic diffraction MTF for a 300-mm EFL *f*/5.4 apochromatic doublet, designed to cover a broad spectral band in the near-IR region.

```
*PARAXIAL CONSTANTS
    Effective focal length:    300.04186    Lateral magnification:   -6.4228e-29
    Numerical aperture:          0.09259    Gaussian image height:      0.00524
    Working F-number:            5.40000    Petzval radius:          -438.15424


*LENS DATA
APO DOUBLET (FINAL)
  SRF        RADIUS        THICKNESS      APERTURE RADIUS          GLASS


  OBJ          --          4.6715e+30      8.1533e+25              AIR


  AST          --           1.00000        27.78165 AS            AIR


   2      156.00000 V       5.00000        30.00000           O_S-NSL36
   3       53.98000 V       0.20000        28.00000              AIR


   4       54.16000 V      14.00000        30.00000           O_S-FPL51
   5    -1165.50000 V     288.81000        30.00000              AIR


  IMAGE        --             --            5.00000
```

```
*WAVELENGTHS
CURRENT   WV1/WW1      WV2/WW2      WV3/WW3
   1      0.85000      0.70000      1.00000
          1.00000      0.77000      0.80000
```

```
*REFRACTIVE INDICES
  SRF     GLASS            RN1          RN2          RN3         VNBR


   0      AIR            1.00000      1.00000      1.00000        --
   1      AIR            1.00000      1.00000      1.00000        --
   2      O_S-NSL36      1.50929      1.51296      1.50676     82.10683
   3      AIR            1.00000      1.00000      1.00000        --
   4      O_S-FPL51      1.49186      1.49420      1.49025    124.37978
   5      AIR            1.00000      1.00000      1.00000        --
   6      IMAGE SURFACE
```

Figure 11.9 Tabulated lens data for the 300-mm EFL apochromatic doublet, including all nominal information required for generating manufacture and assembly drawings.

When the optical system is as simple as this doublet lens assembly, then the optical schematic, lens details, and tolerance data can easily and conveniently be combined into a single document. The drawing shown in Fig. 11.10 contains all information relevant to the manufacture of the apochromatic doublet, including tolerances and a table of basic performance characteristics. The ultimate test of such a document is its completeness (it must have all the data required) and its accuracy (that data must have been transposed correctly).

All tolerances should be evaluated in terms of their impact on image quality. One example of the tolerancing process will be given here to demonstrate this basic approach. The performance of the doublet lens will be degraded when one of the elements is tilted relative to the other. The amount of allowable tilt must be determined and translated to the parallelism, or wedge tolerance, on the lens spacer. Standard machining on a part of this type will easily produce parallelism within 0.001 in. Converting this to an angular tilt of the second element (0.024 deg), it is found that the nominal spot size of 0.006 mm increases to 0.015 mm. This is nearly three times the size of the Airy disk, which is clearly not acceptable. Decreasing the angle between the elements to a tight but achievable tolerance of 0.0002 in. (0.005 deg) results in a spot size of 0.0065 mm. Since the Airy disk size is 0.0053 mm, this amount of degradation is deemed acceptable. Other tolerances on parameters such as thickness, radius, and wedge can be evaluated and determined in a similar fashion. For a simple design such as this it is reasonable to assume the tolerances will all accumulate in an additive, or worst-case fashion. In reality, the probability is great that the actual lens assembly image quality degradation due to manufacturing tolerances will be between one half and one quarter of the worst-case prediction.

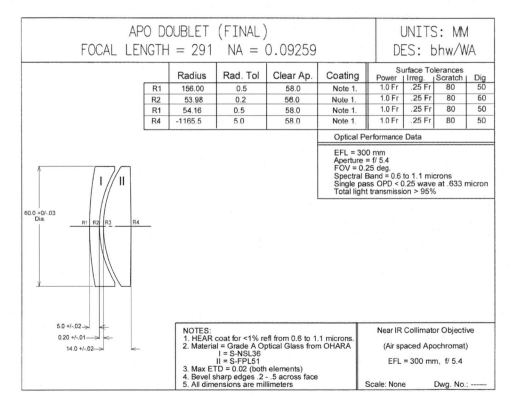

Figure 11.10 Detail drawing of near-IR apochromatic doublet, including all tolerance and performance data.

11.5 Precision Imager Lens Design

11.5.1 Defining the problem

In one family of optical systems, the lens assembly that produces the final image on the detector is referred to as an imager lens. The system that will use the imager lens to be described here consists of an afocal unit (telescope), a beamsplitter assembly, the imager lens, and a solid-state charge-coupled device (CCD) television camera as a detector. All system components have been designed and are in place with the exception of the imager lens.

System requirements dictate the following imager lens specifications:

EFL: 80 mm
Speed: *f*/1.75
Image size: 9.6 mm^2 (design for 13.4 mm diameter)
Spectral band: monochromatic, $\lambda = 0.633$ µm
Entrance pupil location: 60 mm in front of the first element
Vignetting: 10% maximum
Pixel size: 0.02 mm^2 for 80% energy (all field points)

11.5.2 Selecting a starting point

Here again, at this stage of the design process it is wise for the lens designer to consider all the lens requirements and to prioritize them in terms of how each might impact the basic lens design. In this case, the lens focal length and speed are not unusual. The field of view to be covered is reasonable at about ±5 deg maximum. The fact that this will be a monochromatic design will greatly simplify it. The requirement that the entrance pupil be located in front of the lens will make this design somewhat unique and more difficult. The image quality specification is not severe, but the need for uniformity across the entire field, coupled with the relatively small vignetting allowance, will make the design a challenge.

It is also a good idea at the start of each lens design to compute the Airy disk so that the image quality requirement can be compared with it. In this case, the diffraction-limited blur circle radius (the Airy disk) for an *f*/1.75 lens at $\lambda = 0.633$ µm, will be

$$R = 1.22 \times \lambda \times f/\# = 1.22 \times 0.000633 \times 1.75 = 0.0014 \text{ mm}.$$

Having a spot-size requirement (0.02 mm) that is nearly 15× the size of the Airy disk means that all design and analysis work can be conducted on a geometric basis, and diffraction effects will not need to be considered.

The starting lens configuration was chosen after several alternatives were considered and rejected. The Cooke triplet and double-Gauss lens forms offered great promise in terms of lens speed, field of view, and image quality. However, the performance of both of these designs relies greatly on their basic symmetry about a central aperture stop. For the imager lens, the entrance pupil is the aperture stop and, because it is located in front of the lens, that basic symmetry does not exist. A lens type that is frequently used with a forward entrance pupil is the Petzval type. Thus, it was concluded that a derivative of the basic Petzval lens form would offer a good probability of success in this case. A search of available literature turned up a four-element Petzval lens design patent that was selected as a starting point for this design.

11.5.3 Modifying the starting point

The starting Petzval lens design was then modified to conform to the basic requirements of focal length, lens speed, and field of view that were established. The optical glass types were simplified; in fact all elements were changed to BK7 to reduce cost and complexity (it was felt that the monochromatic nature of the design would permit this). It can be seen from the initial patent lens form, shown in Fig. 11.11(a), that the upper ray of the off-axis bundle is being refracted through some rather severe angles between the last two lens elements. The steep inside curves of the rear doublet are obviously not desirable. It was surmised that these curves were involved in correcting color and field curvature, and coma and astigmatism in the original design. Since this is to be a monochromatic design, it was decided to try increasing the doublet lens separation such that the second element becomes a field flattener, located close to the image plane and dedicated primarily to the correction of field curvature. This modified form of the starting Petzval lens was then subjected to a preliminary optimization, with encouraging results. That modified version of the starting Petzval design is shown in Fig. 11.11(b).

This is a logical point in the lens design process to stop and review the design that is evolving in terms of its general configuration. In this case, for example, it was confirmed that the presence of a field-flattener lens in close proximity to the detector would not be a problem. The overall physical size of the lens (length and diameter) can be estimated with some certainty at this point. It should be

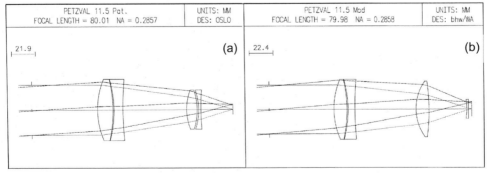

Figure 11.11 The requirements established for the imager lens indicate that a Petzval lens form will be required. A patent was found (a) that will serve well as a starting point for the design. That form was modified and simplified as shown (b). This was the lens form that was input for the final optimization process.

confirmed (with the customer) that the size and weight of the proposed lens will be acceptable in the ultimate application. A preliminary estimate of the approximate cost to manufacture the lens is also possible at this time, on the basis of the modified starting lens form. Having found no apparent pitfalls, the final design optimization is ready to begin.

11.5.4 Optimizing the modified starting design

The final optimization process involves the manipulation of available variables while monitoring the performance characteristics of the lens—particularly image quality. For the imager lens, all surface curvatures and the rear airspace were used as variables. The constraints assigned held the focal length to 80 mm and the speed to *f*/1.75, while minimizing the spot size for three field angles (0, 3.5, and 5 deg). The optimization process depends on an effective software package, combined with its intelligent application by a skilled and experienced lens designer. It is a rare occasion when the designer can just "dump" all requirements into the computer and come up with a successful design with no further intervention. After several cycles of this design process, an optimized configuration for the imager lens was generated. Figure 11.12 contains data describing the starting and ending configurations for this lens. The reduced spot size that has been achieved for all field points indicates that an acceptable design has been generated. More thorough image evaluation will be done in the next stage of the design process to confirm that this is indeed the case.

11.5.5 Evaluating the design

As with earlier examples, the design evaluation covers two very basic and equally important areas. First, the basic lens parameters, including manufacturing tolerances and other aspects of producibility, must be examined. Second, the design's performance in terms of image quality must be evaluated. The design must meet all requirements and expectations in both of these areas in order to be considered a successful design.

The final imager lens design has a focal length of 80 mm and a speed of *f*/1.75, and it produces the required image quality over the entire image. The clear apertures of the lens elements were carefully assigned to assure that the vignetting specification would be met. All lens elements are made from a single common glass type which was selected for its many desirable properties. A quick look at tolerance sensitivities indicates that the design should be quite reasonable as far as manufacture and assembly costs are concerned.

The image quality, in terms of spot size, was constantly monitored during the design optimization phase. Now is the time for a more rigorous analysis. First, the ray-trace analysis will tell the lens designer a great deal about the residual aberrations in the design and how the various aberrations have been balanced to produce the reported spot size. Figure 11.13 shows the final aberration curves for this lens at the three field positions (center, edge, and corner). These curves give

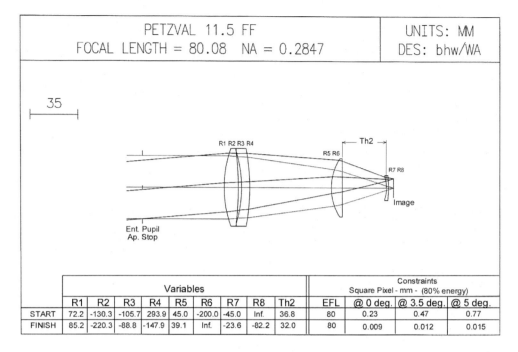

Figure 11.12 The optical layout is shown for the finished imager lens with the variables and constraints that were used during optimization. Values are shown for both the start and finish of the optimization process.

a good indication of the uniformity of performance over the full field of view. Shown elsewhere in Fig. 11.13 are the field curves for astigmatism and distortion. The flat field due to the field-flattener lens is responsible in large part for the uniform image quality. The maximum distortion of <0.5% in the corners of the square image will be quite acceptable. The lens layout, including on- and off-axis rays, is helpful to the lens designer in evaluating the function and significance of the individual components.

Since the original specification of image quality is given in the form of spot size (square pixel), the preferred evaluation for image quality of this lens is done using ensquared energy diagram analysis. Figure 11.14 shows the ensquared energy diagrams for the three field positions (center, edge, and corners). Finally, these ensquared energy curves allow the designer to confirm that the specification for 80% of the energy to be within a 0.02-mm^2 pixel has been met comfortably at all field positions (actual values range from 0.009 at the center to 0.015 in the corners). Maximum vignetting, also in the corners of the image, is found to be 7%.

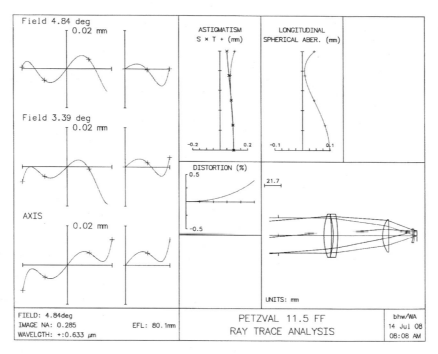

Figure 11.13 Ray-trace analysis of the optimized imager lens design. This basic output from the OSLO optical design package shows all aberration data, along with a convenient lens layout.

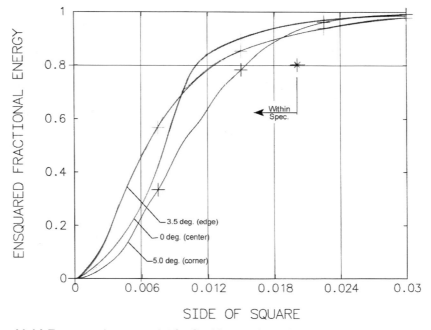

Figure 11.14 Ensquared energy plot for final imager lens design. Results show that 0.02-mm^2 pixels will contain more than 80% energy for all field points within the 9.6-mm^2 sensor.

```
*PARAXIAL CONSTANTS
    Effective focal length:      80.08294    Lateral magnification:   -2.6694e-29
    Numerical aperture:           0.28570    Gaussian image height:     6.78000
    Working F-number:             1.75000    Petzval radius:         -411.71030


*LENS DATA
PETZVAL 11.5 FF
  SRF      RADIUS       THICKNESS    APERTURE RADIUS       GLASS

  OBJ        --       3.0000e+30    2.5399e+29            AIR

  AST        --        60.00000     22.80000 AS          AIR

   2     85.20000 V    10.00000     28.00000 K           BK7
   3   -220.30000 V     2.35193 S   28.00000 K           AIR

   4    -88.80000 V     6.00000     26.00000 K           BK7
   5   -147.90000 V    60.00000     28.00000 K           AIR

   6     39.10000 V     8.00000     21.00000 K           BK7
   7        --    V    32.00000     21.00000 K           AIR

   8    -23.60000 V     1.50000      8.00000             BK7
   9    -82.20000 V     3.84000      9.00000             AIR

  IMS        --           --    V    6.80000


*WAVELENGTHS
CURRENT   WV1/WW1
    1     0.63300
          1.00000


*REFRACTIVE INDICES
  SRF    GLASS           RN1         TCE

   0     AIR           1.00000       --
   1     AIR           1.00000    236.00000
   2     BK7           1.51508     71.00000
   3     AIR           1.00000    236.00000
   4     BK7           1.51508     71.00000
   5     AIR           1.00000    236.00000
   6     BK7           1.51508     71.00000
   7     AIR           1.00000    236.00000
   8     BK7           1.51508     71.00000
   9     AIR           1.00000    236.00000
  10     IMAGE SURFACE
```

Figure 11.15 Tabulated lens data for the imager lens, including all nominal information required for generating manufacture and assembly drawings.

11.5.6 Documenting the final design

Figure 11.15 is a table of lens data for the final imager lens design. As this is a compound lens assembly, the design calls for an optical schematic, followed by a set of detail lens drawings, one for each optical element (as shown in Chapter 7,Fig. 7.5). The optical schematic for this lens is shown in Fig. 11.16. In a lens of this complexity it will be most effective to add to the schematic a separate sheet containing a tabulation of all tolerance data relative to the assembly and alignment of the lens, along with a complete set of detail drawings for each element. These two steps will not be a part of this presentation. The function of Fig. 11.16 is to illustrate the nominal lens configuration and document the basic optical characteristics of the lens assembly.

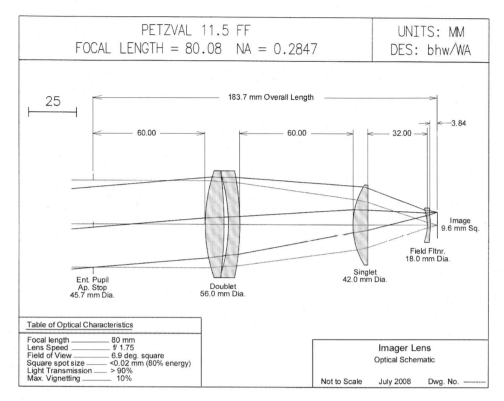

Figure 11.16 Optical schematic of the imager lens showing general information that might typically be included. Comprehensive tolerance data would be contained on a second sheet of the drawing. Optical elements would be described in separate detail lens drawings.

11.6 Unusual Lens Design Task

In nearly every case, it is the lens designer's task to generate a design that has well-corrected image quality. There is one unique category where the designer departs from this philosophy and instead generates a design with very carefully controlled image quality that is sometimes far from the usual goal of perfection. This case arises when designing a null lens for the fabrication and test of an aspheric surface. The following examples will illustrate the procedures used to generate two typical null lens designs.

First, consider the case where a primary mirror is to be produced for use in an IR telescope assembly. This mirror has a diameter of 260 mm, a vertex radius of 355 mm (concave), and a conic constant of –1.0828. A mirror with a conic constant of 0 will be spherical. When its conic constant is –1.00, it is a paraboloid, and when it is less (more negative) than –1.00, as in this case, the mirror surface is a hyperboloid. An ideal method for testing a concave reflecting surface is to generate a monochromatic wavefront that reflects from that nominal surface in a predictable fashion, and then to monitor that reflected wavefront, adjusting the shape of the surface under test until it performs precisely as predicted. When the surface under test is spherical, the center of the projected wavefront is set to coincide with the center of curvature of the mirror and then, when the mirror surface has been figured to the correct (spherical) shape, the wavefront will be reflected precisely back on itself. This reflected wavefront can then be monitored interferometrically and the mirror surface corrected to a small fraction of a wavelength.

If the mirror under test is a paraboloid, then the center of the projected wavefront is placed at the focus of the mirror. The reflected beam will then take the form of a plano wavefront which can be reflected from a flat reference mirror, once again from the mirror under test, and then evaluated interferometrically.

Having a conic constant very close to –1.0, the mirror to be tested in this example is very nearly a paraboloid. If we wanted to test this mirror as a paraboloid, we would assume a point source at its focus and evaluate the error that exists in the reflected beam, which would be perfectly flat for a paraboloid. When the hyperboloid in question is thus evaluated, it is found that there is an OPD error of many wavelengths in the reflected wavefront. This error would not be acceptable in this case since we are striving for a test configuration with less than 1λ of error ($\lambda = 0.633$ μm).

Further analysis reveals that this hyperboloidal mirror has two foci, one at a point 174 mm in front of the mirror and a second that is located 8749 mm behind the mirror. The significance of these foci is that a wavefront originating at one of them will be focused by the mirror at the second, with zero wavefront error. Knowing this, we can conclude that if we place our interferometric point source at the first focus, 174 mm in front of the mirror under test, the reflected wavefront from a correctly finished mirror will be perfectly spherical and it will appear to be coming from the second focus which is located 8749 mm behind the mirror surface. This condition is illustrated in Fig. 11.17. It can be seen that if a spherical test mirror is placed at the proper location, the matching spherical wavefront will be reflected back on itself such that it precisely retraces its original path. With a test setup as shown, any error in the returning wavefront will be a precise indication of the error that exists in the hyperboloidal mirror under test. The spherical test mirror in this configuration is also known as a *Hindle sphere*.

The Hindle sphere is one of the simplest and most basic forms of null lens. More typically, the aspheric surface will be more complex and the null lens design will be generated using a somewhat different approach. For example, the same IR telescope design that uses the primary mirror just described contains a refracting surface with a concave aspheric shape in the form of an oblate spheroid, with a vertex radius of 31.84 mm, a diameter of 28 mm, and a conic constant of +0.0806. The approach to designing a null lens to test this surface will be to generate a refracting lens assembly that will produce a diverging wavefront that precisely matches the shape of the desired aspheric surface.

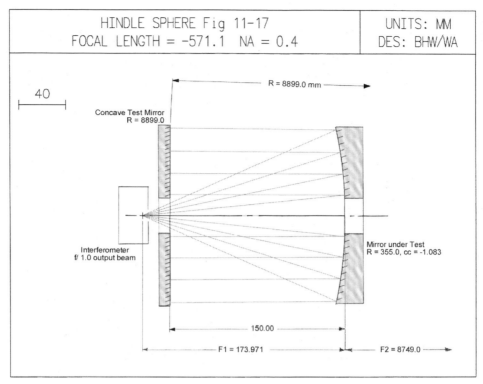

Figure 11.17 Test setup for manufacture and test of a concave hyperbolic primary mirror. An interferometer point source is located at the first focus (F1) of the mirror under test. The reflected wavefront is spherical, appearing to come from F2. The concave test mirror (Hindle mirror) reflects this spherical wavefront back to its origin within the interferometer where errors in the mirror under test are observed.

The degree of departure from spherical of the surface under test will be an indicator of the complexity that will be encountered in the null lens design. In this case, the surface asphericity is greater than five waves. Past experience indicates that the null lens design for manufacture and test of a surface having this much asphericity will take the form of a two-element lens assembly. Once the basic lens form is selected, the design is then optimized, using all null lens curvatures and spacings as variables. Figure 11.18 shows the resulting arrangement, including the prescription of the null lens design. As with the Hindle sphere, the surface test is accomplished by introducing an interferometric point source which creates a perfect spherical diverging wavefront. In this case, the null lens then introduces additional divergence along with wavefront aberration that results in the transmitted wavefront precisely matching the shape of the surface under test. In terms of ray tracing, when the surface has reached the required shape, each ray emerging from the null lens will strike the surface under test at exactly 0 deg angle of incidence. The result is that each ray will then be reflected precisely back on itself, retracing its path through the null lens and

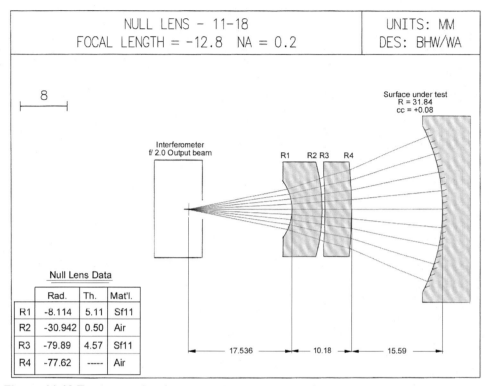

Figure 11.18 Test setup showing a two-element null lens for the manufacture and test of a concave aspheric surface. With the point source located as shown, the wavefront emerging from the null lens will be shaped to exactly match the aspheric surface. Interferometer output will reveal any error in the shape of the surface under test.

back to the point source. The shape of the reflected wavefront can then be precisely evaluated interferometrically, with any error in its sphericity an indication of residual error in the surface under test.

These two examples of null lens design have been introduced to give the reader some feel for the diversity of work that might typically be encountered by the lens designer. The null lens falls into the category of test equipment and as such must be handled differently when it comes to tolerancing and documentation. The key point here is that the null lens will not, in most cases, be delivered to the customer, but will rather become a tool in the inventory of the manufacturer. In general, basic tolerances for test equipment must be held tight enough to assure the performance of the lens being manufactured. Often, null lens tolerances will be tighter than those on the part being manufactured by a factor of 10×. On the other hand, cosmetic defects such as edge chips and scratches, and certain physical tolerances such as lens diameter, will not have to be held as tightly. An additional economic consideration is that a good null lens assembly will probably serve for the manufacture of many parts, thus its initial cost can be spread out over the life of a contract involving many repeat uses.

11.7 Review and Summary

This chapter has been presented to introduce the reader to the field of lens design and to demonstrate a few of the unique and interesting aspects of the field of optical engineering. Three essential factors are required to create a successful design: a skilled and experienced lens designer, a well-executed optical design software package (program), and a powerful and fast computer. While all three are essential, they have been listed in order of importance.

While the lens design process must remain flexible, a typical process has been outlined here that will serve well as a guide to one entering the field. Several examples have been presented to demonstrate the methods that might typically be employed in generating a lens design using the suggested typical lens design process. The final section deals with the design of null optics for the fabrication and testing of aspheric surfaces. This has been included to demonstrate the wide range of design tasks that the lens designer might be called on to execute. In this case, the typical procedure is departed from in order to meet the rather unusual requirements.

All the work done in this chapter has been done using the OSLO Premium Edition, Rev. 6.1 optical design program on a Dell XPS410 personal computer. The field of computer hardware and software today is one of the most dynamic ever experienced in the history of technological development. While many changes will continue to occur during the lifetime of this edition, it is believed that this chapter will maintain its value as a tool to help the reader to understand and (hopefully) enjoy the subject of lens design.

Chapter 12
Optics in Our World

12.1 Introduction

Throughout this book we have discussed many facets of the field of optics and optical engineering. It is my hope that this exposure will encourage some readers to further pursuits and studies, perhaps leading to a career in this field. For others this has merely been an excursion into an area of general interest and nothing more. I believe this final chapter will be of interest to all of these readers. It will deal with the optical aspects of several phenomena and devices that we all might encounter in our daily lives. It is my hope that this will increase our awareness, appreciation, and understanding of them all.

We have seen in several earlier chapters how the human eye is closely related to the optical design of certain instruments. This chapter will touch on the concept of the complete visual system, including the eyes, the mind, and our lifetime of visual experiences. These experiences tend to program the mind, allowing it to interpret what is seen by the eye and imaged onto the retina.

Also in this chapter, we will deal with a number of technological advances of the day, with special emphasis on their optical content. In order to keep this presentation at a reasonable comfort level (for the author as well as the reader), this section will be limited to the most fundamental concepts and principles.

12.2 Optical Illusions: Size

Many times what we actually see, and what we think we see, are two very different matters. This phenomenon results when the mind, based on earlier experience, adds information to what is collected and delivered by the eye, leading us to a conclusion that may not always be correct.

In Fig. 12.1, for example, the diagram on the left represents a common optical illusion known as the Ponzo illusion. Here we see two rectangular blocks (A and B) that are exactly the same size. It follows that the images of blocks A and B on our retina are also the same size. In spite of these facts, the other clues included in the figure lead us to conclude that the upper block (B) is substantially

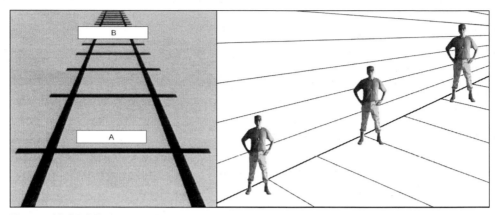

Figure 12.1 While the eye is a precise optical device, the image produced on the retina is interpreted by the mind where other visual clues are introduced which sometimes fool the viewer into reaching the wrong conclusion. For example, is block A on the left really smaller than block B? Are the three soldiers in the figure on the right really different heights? Go ahead . . . use a ruler to check it out.

larger than the lower block (A). The deception begins when our mind relates the two converging vertical lines to a lifetime of experience dealing with conditions similar to standing and looking down a long, straight highway or stretch of railroad tracks. On the basis of this experience, we conclude that the two lines in the figure are actually parallel and that they appear to converge only because the tops of the lines are much farther from us than the bottoms. It is a small leap then to conclude that the upper block is also farther away from us than the bottom block. Now the mind goes into action: we see two blocks that appear to be the same size, but believe one to be farther away from us than the other; therefore, obviously the one that is farther away must be larger. Two different size blocks is not what our eye sees, but rather it is what our mind perceives. This is an optical illusion.

The diagram on the right in Fig. 12.1 is a similar illusion. The perspective layout indicates a level surface and a vertical wall, both apparently receding from us. The three soldiers shown are all the same height. Because other clues indicated they are at different distances from us, we conclude that they are progressively larger as their distance from us increases. For most viewers, both of these illusions will be slightly enhanced when viewed with one eye rather than two. This happens because when we view an object with both eyes, the eyes converge to a point on that object. In the case of the three soldiers, this clue (the amount of eye convergence) tells our mind that all three are at the same distance. This tends to offset, to a small degree, the other clues that are indicating otherwise. Viewing with just one eye removes the eye convergence clue from the set of information being processed by the mind.

The moon illusion is one that has been witnessed by most of us. When we see a full moon rising on the horizon, it is generally perceived to be much larger than when it is seen later the same night, more nearly overhead. The basis for this

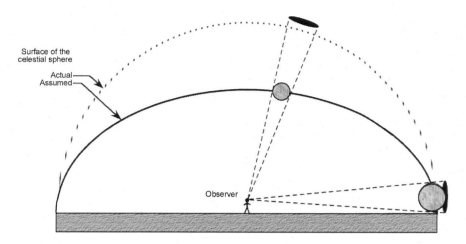

Figure 12.2 The moon illusion causes the perceived size of the moon to change depending on its location in the sky. When the moon is near the horizon, we erroneously assume it to be at a greater distance than later at night when it is more nearly overhead. Because we assume the celestial sphere to be flattened, we assume that as the night progresses, and the moon moves away from the horizon, it is also moving closer to us. Like the blocks A and B in Fig. 12.1, the size of the moon is the same in both cases. If the moon appears the same size, in spite of its having (apparently) moved closer to us, we can only conclude that it has grown smaller as it has moved from the horizon to the near overhead position.

illusion is quite similar to those just described and illustrated in Fig. 12.1. The concept of the celestial sphere involves an imaginary surface on which all the stars, planets, and other heavenly bodies are located. If we stand or lie in the middle of a large flat expanse, on a clear night, it is not difficult to imagine such a surface. For most of us, as with the observer in Fig. 12.2, we perceive this celestial sphere to be flattened considerably relative to a true hemisphere. This may be due to our lifetime of exposure to cloud-filled skies during the daylight hours, where those clouds near the horizon are considerably farther away from us than those that are nearly overhead. In any event, as we gaze at the star-filled night sky, it is quite likely that the surface containing all the stars, planets, etc., will be perceived as having the shape shown in Fig. 12.2. As the moon rises above the horizon, it is assumed to be on that surface and its size is judged accordingly. As the night progresses, and the moon appears more nearly overhead, our presumptions tell us that it is now closer to us. Because its angular size has remained constant, in spite of having (apparently) moved closer to us, we can only conclude that the moon has become smaller than it was when it was on the horizon. Again, our mind leads us to a conclusion that contradicts what our eye sees and what our intelligence tells us is actually the case. As in Fig. 12.1, making actual measurements of the blocks and soldiers to confirm their size does not cause the illusion to cease, just as knowing that the size of the moon is constant does not diminish the effect of the illusion. In these examples, and with the moon illusion, what we see, what we know, and what we think, are passed to the brain and processed to generate the ultimate perception of the mind.

12.3 Other Optical Illusions

Optical illusions such as the moon illusion relate to our judgment of the size of certain objects. Other illusions deal with the perception of shapes and objects that might not otherwise actually exist. For example, the diagram on the left in Fig. 12.3 at first appears to be just a random collection of straight lines of different lengths, some horizontal, some vertical. The significance of the pattern is not obvious (to most of us) on the basis of our experience viewing similar patterns. When it is suggested that this might be a block letter **H** that is being illuminated from the upper left, our mind accepts that premise and the complete outline of the letter is now perceived. It will now be difficult to view this pattern of lines ever again without seeing the block letter **H**.

In the center of Fig. 12.3 we can see a rectangular block with a diagonal line (A) passing behind it from left to right. Emerging on the right are two parallel lines, one of which is a true extension of line A. The interruption of the line, combined with its diagonal orientation, leads most to conclude that line B is the extension of line A. If we lay a straight edge on the figure, it is apparent that our mind has again been fooled—it is line C that is the extension of line A. Interestingly, this is another case where being made aware of the reality does not cause the illusion to go away.

While the letter **H** was apparent only after we had been given a suggestion of its existence, the diagram on the right in Fig. 12.3 requires no such clue. Here we have a pattern of incomplete circles from which a white triangle literally leaps out at us, in spite of the fact that no white triangle actually exists. Our mind leads us to conclude that the only reason these circles would appear as they do is because this white triangle has been placed between our eyes and the circles. So, the nonexistent white triangle is clearly perceived by our mind's eye.

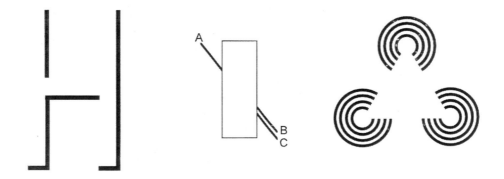

Figure 12.3 Shown here are three common optical illusions. The pattern on the left appears, at first, to be a random set of straight lines, until it is suggested that this might represent a block letter H, illuminated from the upper left. The lines are then interpreted as shadows of the letter edges, and the letter H is then easily perceived. In the center figure, the question is: which line, B or C is an extension of line A? On the right, the arrangement of interrupted circles leads to the perception of a white triangle, in spite of the fact that there are no lines present to indicate the actual outline of that triangle.

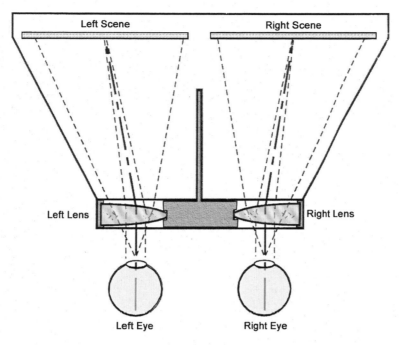

Figure 12.4 The stereoscope is a viewing device that produces a three-dimensional effect when it is used to view a properly prepared stereo pair of pictures. The lens segments introduce both lens power and prism power, which allow the eyes to relax both focus and convergence. The final and most important function of the stereoscope is to direct the left scene to the left eye and the right scene to the right eye.

12.4 Seeing the Third Dimension

As we view the outside world, our visual system easily perceives the three dimensions of width, height, and depth. When we view the printed page, the sensation of depth is greatly diminished. When we are viewing printed reproductions of the outside world in the form of drawings and photographs, there are generally enough clues present to allow us to perceive depth where it does not really exist. Methods exist that allow us to maximize and even enhance the third dimension of depth when actually viewing a two-dimensional object. In the mid-1800s the stereoscope became a familiar and popular method of home entertainment. Utilizing the basic optical arrangement shown in Fig. 12.4, the stereoscope simultaneously accomplishes three functions, each adding to the three-dimensional illusion. First, the lens power that is present allows the viewer to relax the focus of the eyes, which indicates to the mind that the scene being viewed is at a great distance. Second, the prism effect of the lens segments allows the viewer to relax the convergence of the eyes, again indicating a distant scene. Finally, and most important, the stereoscope is designed to view a pair of photos of the same object that have been recorded from slightly different points of view, much as the object would have been seen by the viewer's two eyes. The stereoscope is configured such that each eye sees one of these photos and not the other. The result is the eyes focus and convergence are relaxed, while each eye sees a view of the scene from a slightly different point of view. When this information is processed in the viewer's mind, the illusion of a true three-dimensional scene is realized. In modern time the original stereoscope has been repackaged into the familiar Viewmaster, an educational toy used for viewing a variety of cartoon characters as well as real-world scenes by our children. Today, more sophisticated high-performance versions of the stereoscope and the Viewmaster are available, which present scenes with greatly improved image quality and increased apparent field of view for truly exciting three-dimensional viewing.

A pseudostereo effect will be observed by most viewers by simply viewing a printed image with one eye while the other eye remains open and relaxed, but covered. This effect is most pronounced when viewing a large picture of a scene containing numerous clues relating to depth. An object such as a road, river, or fence that extends from close to the point of observation in the foreground, deep into the scene, will help reinforce the illusion. The result of covering one eye will generally be that the convergence of the eyes will relax and the axes of the two eyes will become nearly parallel. This removes the convergence clue from the process and allows the mind to accept the fact that objects in the scene may be at varying distances, based on their appearance and size. The illusion of depth that will be achieved depends on the relaxed state of the viewer's eye and the content of the image being viewed. Once achieved, it is interesting to uncover the second eye and note how the depth effect immediately ceases to exist following the reintroduction of the convergence clue.

Three-dimensional optical systems have come and gone from the entertainment scene over the years. The three-dimensional movie came on the scene in the 1950s, but failed to catch on. In large part this was due to the advent of several spectacular wide-screen processes (CinemaScope, Cinerama, Todd AO, etc). It was found that the added benefits of a wide-screen presentation were better received than the somewhat artificial effects of three-dimensional viewing. In more recent years, the advantages of three-dimensional have come to the fields of tactical reconnaissance and have proven to be very worthwhile. In another related development, the endoscope, used in noninvasive surgical procedures, has recently been modified to provide a true three-dimensional view to the surgeon. The advantages in this application have been described as very significant in the success and effectiveness of such surgeries. While the potential for an in-home three-dimensional high-definition television system does exist, it appears at this time that, as it was with the motion picture in the 1950s, the large-screen television approach will dominate the market for a few more years.

12.5 Optics and the Compact Disk

Many of the optically related technological advances of today have impacted the field of home entertainment. By the 1990s, the compact disk (CD) and the digital video disk (DVD) replaced the phonograph record, audio tapes, and video tapes in most home entertainment systems. This has been made possible by simultaneous advances in laser systems and data-processing electronics. The phonograph record, in spite of many improvements over the years, has only ever been a molded spiral groove in a flat disk. With output from a microphone used to record the sound, the configuration of the groove is physically altered. On playback, the needle (stylus), tracking in the groove of the record, will vibrate in a way that is analogous to the sound waves that were recorded. The mechanical motion of the stylus is then converted to an electrical signal, which is amplified and processed to drive a loud speaker, thus reproducing the original sound.

All this has been changed by the ability to produce a comparable digital electronic signal that contains all the information needed to reproduce the recorded sound. In the digital recording system the output from the microphone is converted into a continuous digital signal. This signal is then converted to the binary form, much like what is used in the processing of most computer data. In the binary language, all numerical data are converted to a series of bits, each represented by either a 0 or a 1. In this way the recorded signal can be processed using techniques developed for computer applications. The result is that any piece of recorded music (or other information) can be transformed into a long string of 0s and 1s. This string of binary digital data is then transferred to a spiral track on a CD, where one of the binary digits is made highly reflective while the other is made to absorb incident energy (see Fig. 12.5). Now, the optical aspect of the CD system comes into play. The CD is played back while being rotated at a variable rate, such that the linear speed of the data track is constant. Energy from a solid-state laser source is collected by a precision lens assembly and focused down onto the data track. The reflected laser energy is collected and analyzed, resulting in a digital electronic signal that precisely duplicates the original recorded data. The accuracy of the digital process is enhanced greatly by the fact that the playback process is noncontact; thus, wear and tear on the CD is minimal and its life is essentially infinite. Fig. 12.5 illustrates the basic principle of the CD and the CD playback mechanism.

A similar approach to the playback of video signals is incorporated into the digital video disk (DVD) system, where both pictures and sound are contained on a disk similar to the CD. In most home entertainment systems the CD and the DVD have replaced audio and video magnetic tapes. In addition to the advantages of increased fidelity and a reduced wear factor, the quick access to random locations within the recorded program is greatly enhanced by the digital approach.

12.6 Optics and Projection TV

The most common form of television set in home use today is referred to as *flat screen, LCD,* or *plasma.* In the flat-screen set the picture is generated electronically by a matrix of picture elements (pixels), and the viewer observes that output directly. While the liquid crystal display (LCD) or plasma flat screen is the system of choice based on overall image quality, this approach does limit the size of the display that can be reasonably produced and used in the home. In the quest for larger displays at reasonable cost, several projection TV schemes

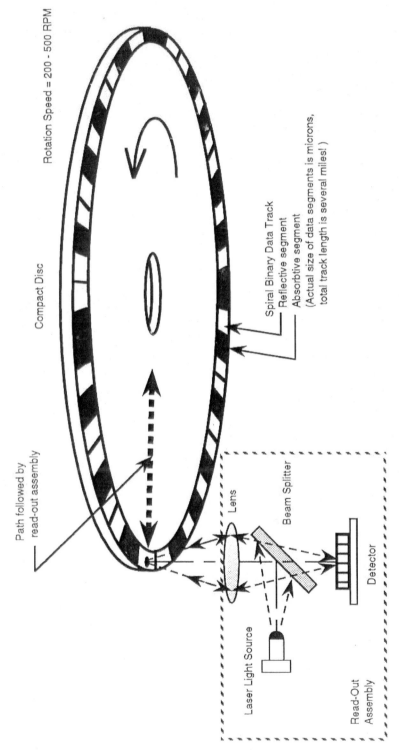

Figure 12.5 The compact disk (CD) playback mechanism includes a precision turntable functioning in conjunction with a read-out assembly that contains a solid-state laser light source, a beamsplitter, a precision lens assembly, and a detector that senses the presence or absence of a signal reflected from the data track on the disk.

have been developed that offer quite good performance and considerable promise for the future. All of these systems involve rather complex and sophisticated optics.

The two basic approaches to projection TV are the *front-projection* and *rear-projection* systems. The names are derived from which side of the viewing screen the picture is projected on. As is usually the case there are pros and cons involved with either approach. Figure 12.6 shows schematically the typical optical arrangement used for both front- and rear-projection TV systems. The front-projection system shown uses three sources which generate identical pictures, except for their color content. Beamsplitters are used to combine the three pictures which are then projected onto the front surface of a large reflective screen. As with the conventional motion picture, the projected scene is reflected from the screen for viewing by the audience. This front projection approach offers a compact projection unit and considerable flexibility with regard to the ultimate size of the projected image. On the down side, the front-projection system is made up of two separate units (projector and screen) and requires a clear line of sight between the two.

The rear projection system is somewhat more bulky, but it is a self-contained unit, offering more flexibility in terms of its location within the viewing space. In this case the picture is projected onto the rear surface of the screen and it is seen by the viewer as it is transmitted (and slightly dispersed) by the screen material.

As programming sources become more numerous and the TV system continues to merge with computer and video game systems, it becomes more common to find a dedicated media center as an integral part of the design in many home layouts. As this situation continues to develop, it would seem that the large-screen projection TV system will become more affordable and more commonplace.

12.7 Optics and Photography

As we proceed into the 21st century, the field of amateur photography is undergoing a number of monumental changes. This section will touch on a few of these, particularly as they relate to the subject of optics and optical engineering.

The field of 35-mm film photography has long been the choice format of the serious amateur. This system, along with all film-based systems, has largely been replaced today by the introduction of digital video and still photography systems. Developments in the area of digital photography, or electronic imaging, are coming at a fast and furious pace. The status of electronic imaging will no doubt change significantly before this book reaches the reader. A brief review of the field will serve to illustrate the point.

One of the earliest demonstrations of electronic imaging that impacted the masses was the development of broadcast television. Here, a studio camera recorded images and transformed them to electronic signals which were broadcast to the home. In the home, the television set converted the electronic

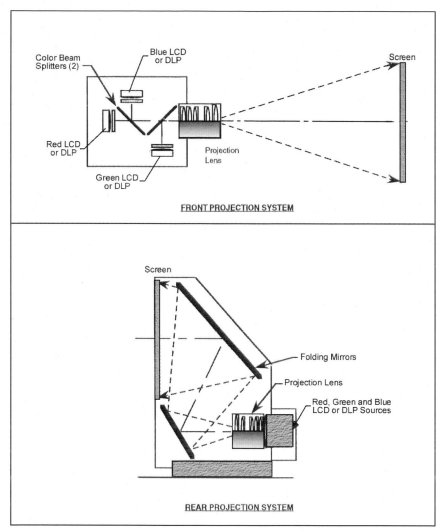

Figure 12.6 The conventional home TV set with a cathode ray tube (CRT) source has been replaced by a family of plasma and LCD flat-screen devices. In addition, in order to achieve even larger picture sizes, several front- and rear-projection designs, such as those shown here, are presently in use.

signal back into a viewable picture on a CRT, or picture tube. In the 1950s and 1960s, more and more homes were equipped with television receivers. The creation and broadcast of TV signals was left (primarily) to the major networks.

During this same time period, if we wished to have motion pictures of our own creation (family, friends, vacations, etc.), these were produced on film, using amateur motion picture cameras, and viewed in the home, using an 8- or 16-mm projector and screen. Likewise, still photos of that time period were recorded on film and viewed at home in the form of prints or projected slides. Professional-quality entertainment was found in the theaters, where one could view motion pictures produced by the major film studios, on the large "silver screen."

In the 1980s, the development of the video camera (camcorder) and the videocassette recorder (VCR) threw the entertainment industry into relative chaos. Today the home movie camera has been essentially displaced by the digital camcorder, while the DVD player has dramatically altered the variety of source material available to the television viewer forever. As the size and picture quality of the television set continues to improve, the motivation for leaving the home to be entertained rapidly decreases. While film continues to be used as the recording medium for most motion picture studios, digital video cameras are rapidly approaching film systems in terms of quality and capability.

In the field of still photography, the digital camera has essentially replaced the film camera for most photographers, both amateur and professional. Figure 12.7 shows a typical compact digital still camera on the left, and a digital video camcorder (camera/recorder) on the right.

This entire topic (electronic imaging) represents an extremely beneficial fusion of multiple technologies. The camera, both still and video, the playback system, the print (or slide), scanners, the video camera, the DVD player, and the modern home computer system have all come together in a spectacular blending of technologies. Optical engineering remains the keystone of the process, with a quality lens responsible for the capture of the original image and its transfer into the electronic domain. Computer hardware and software are major contributing factors to the success of any electronic imaging system, where the potential for image enhancement and manipulation appears to be essentially unlimited.

The ultimate in today's technology is perhaps demonstrated by the CD-ROM (compact disk, read-only memory) system, which brings a vast volume of recorded pictures and sound to the computer for instant access and display. These are indeed exciting times to be involved with the science of optics, as the wave of electronic imaging carries us on into this new century.

(a)   **(b)**

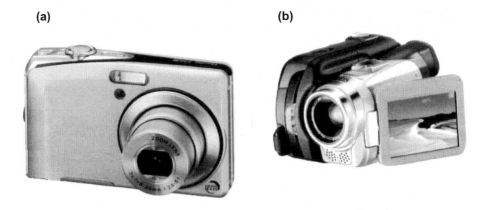

Figure 12.7 Typical compact digital cameras. Point and shoot still camera (a) and digital video camera (b).

12.8 Optics of the Rainbow

Doubtless, one of the most beautiful and spectacular optical phenomena of nature is the rainbow. Here we have, on a most grand scale, a demonstration of the lens, the prism, refraction, reflection, and dispersion, all observed by the human visual system. Figure 12.8 illustrates a particularly spectacular view of a nearly complete primary rainbow, along with the hint of a secondary rainbow. With the sun at the viewers back and rain falling at some distance in front, the sunlight that is incident on the falling raindrops (essentially solid spheres of water) is refracted, dispersed, and reflected back to the viewer. The geometry of the situation results in the red light being bent through an angle that is about 2 deg less than the blue light. The result is that we see an arcing band of light across the sky. The width of that band subtends a 2-deg angle, with its color ranging from red on the outside, yellow and green at its center, to blue on its inside. Under extraordinary conditions, as in this photo, it is sometimes possible to observe a considerably less bright secondary rainbow outside of the primary. The spectral order of this secondary rainbow will be reversed relative to that of the primary because it has been created by an additional reflection within the raindrop.

12.9 Review and Summary

This chapter has been included in the hope that it will convey to the reader some measure of the fun and excitement that can be a part of optics and optical engineering. The discussion of optical illusions illustrates the complexity and sophistication of the human visual system, which includes our built-in camera (the eye), our detector array (the retina), and a data-processing capability beyond that of most computers. A brief discussion of three-dimensional, or stereo, viewing has also been included, with reference to applications in the field of entertainment, along with more serious realms.

Modern technology abounds with developments that are based on the science of optics and the work of the optical engineer. In recent years, home entertainment has been revolutionized by digital recording and optical playback schemes applicable to both sound and video systems. Professional and amateur photography have benefited greatly from recent advances in lens design, camera system design, and newer and better electronic data collection systems. Developments in the field of electronic imaging appear to have led us to that point in time where film, as we know it, is nearly obsolete and the entire field of image recording and viewing is now dominated by digital electronics, computers, and microprocessors.

A basic understanding and appreciation of just a few fundamental optical principles, as presented throughout this book, should serve you well as you prepare to observe and appreciate many of the optical phenomena that may be encountered on a daily basis. And, in the best possible scenario, you will be inspired to explore further this most exciting and rewarding field, perhaps choosing it for your ultimate life's work. As one who has made that passage I can assure you, you will not regret it.

Figure 12.8 Under ideal conditions a full rainbow pattern will be observed. An extremely faint secondary rainbow may also appear, with the order of its colors reversed.

Appendix A
Basic Optical Engineering Library

This addendum contains a listing and description of available textbooks and reference literature that persons working in the field of optical design will find informative and helpful. While deserving of considerable and justifiable debate, the order in which the "Top 10" have been presented reflects my opinion regarding their relative importance and usefulness, based on my experience while working in the field.

1. Warren J. Smith, *Modern Optical Engineering*, 4th ed., McGraw-Hill (2007).
   Since it was first published in 1962, Modern Optical Engineering has become the standard text in the field of optical engineering. It presents a unique combination of the required traditional concepts that have been taught for decades, including the modern ideas and methods employed widely throughout the optics industry today.

2. Rudolph Kingslake, *Applied Optics and Optical Engineering,* Academic Press (1966).
   This series of books deals with optical theory, design, and optical engineering. The first five books in the series are especially useful to the engineer in need of general tutorial and reference information. Subsequent books in the series are dedicated to more specific topics, and they are highly recommended to persons working in those areas.

3. David Falk, Dieter Brill, and David Stork, *Seeing the Light,* Wiley (1986).
   This is a very interesting and informative book dealing with the subjects of physics and light as they are encountered in nature and in science. It covers such topics as geometric optics, photography, vision, physical (wave) optics, and holography. The information on the human visual system, including the perception of color, is particularly well done.

4. *Military Standardization Handbook: Optical Design* (MIL-HDBK-141), Defense Supply Agency, Washington D.C. (October 1965), as appears at: http://www.optics.arizona.edu/ot/opti502/MIL_HDBK_ 141.html.

    First published by the government in the early 1960s, *MIL Handbook 141* has served as a basic reference to optical engineers and designers since that time. The handbook contains a complete treatment of the basic methods employed in the design of optical instruments.

5. *The Photonics Directory*, Laurin Publishing Co. (annual), as appears at: http://www.photonics.com.

    The Photonics Directory consists of four separate volumes: *The Corporate Guide, The Buyers' Guide, The Photonics Handbook,* and *The Photonics Dictionary.* The corporate and buyers' guides offer a comprehensive picture of the photonics industry on an international scale. The handbook and dictionary provide the reader with a wealth of valuable and useful reference information.

6. Rudolph Kingslake, *A History of the Photographic Lens*, Academic Press (1989).

    This book conveys to the reader a sense of the history of the optics industry—especially the history of lens design as it relates to the field of photography. A recognized authority in this field, Dr. Kingslake has contributed significantly to the training of many of today's leading optical engineers while at the University of Rochester.

7. Warren J. Smith, *Modern Lens Design, Second Edition,* McGraw-Hill (2005).

    Primarily a resource manual, *Modern Lens Design* contains some very valuable information dealing with automatic lens design, optimization, and evaluation. This is followed by a comprehensive presentation of complete lens design and aberration data on essentially all lens types. These designs will often serve well as starting points in the generation of new lens designs.

8. Milton Laikin, *Lens Design, Fourth Edition,* Taylor & Francis, Inc. (2006).

    Similar to Modern Lens Design, this book presents a section on lens design methods, followed by complete prescription data on a number of interesting and useful lens designs, with MTF data. Several appendices contain valuable reference information.

9. Francis W. Sears, *Optics*, Addison-Wesley (1949).

    This is a basic text which deals with the fundamental concepts of both physical and geometric optics. Prepared for use in the teaching of general physics, this book serves well as a basic reference. Its vintage precludes any coverage of modern lens design techniques.

10. Jurgen R. Meyer-Arendt, *Introduction to Classical and Modern Optics*, 4th ed., Addison-Wesley (1995).

   This is a very interesting book which deals primarily (and quite nicely) with the subject of physical optics and electromagnetic wave theory. Especially valuable and interesting is the complete and thorough treatment of the history of optics and the key persons involved with the science over the years.

11. Francis A. Jenkins and Harvey E. White, *Fundamentals of Optics, 4th ed.*, McGraw-Hill (2001).

   This is a sweeping treatment of the subject. A section on geometric optics deals with basic components and optical instruments. More time is spent on the subject of physical optics, wave theory, and interference. Finally, a portion of the book presents a discussion of quantum optics and photon theory.

12. Donald H. Jacobs, *Fundamentals of Optical Engineering*, McGraw-Hill (1943).

   This book deals with geometric optics and instrument design. Based on activity during World War II, the presentation is refreshingly basic, and it concentrates on fundamental principles. Fortunately for the reader, much of the information contained is timeless, while the presentation is very clear and easy to follow.

13. Rudolph Kingslake, *Lenses in Photography: The Practical Guide to Optics for Photographers*, A.S. Barnes & Co. (1963).

   This book will be especially useful to the optical engineer involved in the field of photography. Valuable and interesting historic information is presented, along with considerable technical data and reference information difficult to locate elsewhere.

14. Earle B. Brown, *Modern Optics,* Reinhold (1965).

   Modern Optics was written at a time when the optical industry was in a state of general turmoil and rapid growth. As a result, it reflects the status at a most interesting time. General system and instrument design is covered, along with an interesting introduction to the laser and its applications.

15. Philip Kissam, *Optical Tooling,* McGraw-Hill (1962).

   This book deals with the general subject of optical instrumentation and optical tooling in particular. The book was written at a point in time when the subject was peaking in its popularity and applications. It is well written and contains considerable useful information regarding optical tooling instruments and accessories of that time.

16. John Strong, *Concepts of Classical Optics,* Freeman (1958).

    This textbook was written to be used in teaching an intermediate optics course. It is very well written and beautifully illustrated. It covers both physical and geometric optics, with much information relating to the manufacture and test of precision optics.

17. Max Born and Emil Wolf, *Principles of Optics*, 7th *ed.,* Cambridge Univ. Press (1999).

    This is the ultimate text on the subject of electromagnetic theory, interference, and diffraction. The mathematical presentations are rigorous and will be challenging to most readers. This is the definitive book dealing with physical optics.

Appendix B
Optical Design
Software Sources

This addendum contains a listing of the principal sources of optical design software at the date of publication (2008). The vendors appearing here offer software packages that are capable of complete optical design, optimization, and image evaluation. The dynamic nature of this field suggests that the reader consult the most recent reference and trade publications available and the referenced web sites to obtain up-to-date information relative to these companies, as well as possible newcomers to the field.

Breault Research Organization, Inc.
6400 East Grant Rd., Suite 350
Tucson, AZ 85711
Tel: 800-882-5085
www.breault.com

Lambda Research Corp.
25 Porter Rd.
Littleton, MA 01460
Tel: 978-486-0766
www.lambdares.com

Optical Research Associates
3280 E Foothill Blvd., S 300
Pasadena, CA 91107
Tel: 626-795-9101
www.opticalres.com

Sciopt Enterprises
P.O. Box 20637
San Jose, CA 95160
Tel: 408-268-8934
www.sciopt.com

Zemax Development Corporation
3001 112th Avenue NE, Suite 202
Bellevue, WA 98004
Tel: 425-822-3406
www.zemax.com

Appendix C
Optical Glass Sources

Primary Sources

Schott Glass Technologies, Inc.
400 York Ave.
Duryea, PA 18642-2026
Tel: 570-457-7485
www.schott.com

Hoya Optics, Inc.
3400 Edison Way
Fremont, CA 94538-6190
Tel: 510-252-8370
www.hoyaoptics.com

Ohara Corporation
50 Columbia Rd.
Branchburg, NJ 08876
www.oharacorp.com

Sumita Optical Glass, Inc.
www.sumita-opt.co.jp

Secondary Sources

Corning Incorporated
One Riverfront Plaza
Corning, NY 14831
Tel: 607-974-9000
www.corning.com

Newport Glass Works, Ltd.
10564 Fern Ave.
Stanton, CA 90680
Tel: 714-484-7500
www.newportglass.com

Glass Fab, Inc.
P.O. Box 31880
Rochester, NY 14603
Tel: 585-262-4000
www.glassfab.com

United Lens Co., Inc.
259 Worcester St.
Southbridge, MA 01550-1325
Tel: 508-765-5421
www.unitedlens.com

Advanced Glass Industries
P.O. Box 60467
1335 Emerson St.
Rochester, NY 14606
Tel: 585-458-8040
www.advancedglass.net

Appendix D
Conversion Factors and Constants

Acceleration due to gravity	$g = 980.67$ cm/s^2 = 32.174 ft/s^2
Ampere	1 A = 1 coulomb (C)/s
Angstrom	1 Å = 1×10^{-8} cm = 1×10^{-4}μm
Atmosphere	1 atm = 760 mm Hg = 14.696 lb/in^2
	= 1.013 X 10^6 dyn/cm^2 = 1033 g/cm^2
Atomic mass unit	1 amu = 931 MeV
Avogadro's number	$N = 6.022 \times 10^{23}$
Boltzmann/s constant	$k = 1.3807 \times 10^{-16}$ ergs/K
British thermal unit	1 Btu = 252 calories
Calorie	1 cal = 4.184 joules (J)= 0.04129 liter atm
Centimeter	1 cm = 0.01 m = 0.3937 in.
Centimeter/ second	1 cm/s = 0.02237 mi/h
Cubic centimeter	1 cm^3 = 0.06102 in.^{3}
Cubic inch	1 in.^{3} = 16.387 cm^3
Density	$D_{water} = 1.000$ g/cm^3 at 4°C
	$D_{mercury} = 13.6$ g/cm^3 at 0°C
	$D_{air} = 1.293 \times 10^{-3}$ g/cm^3 (at STP)
e (natural log base)	$e = 2.7183$
Electronic charge	$e^- = 4.80 \times 10^{-10}$ esu =1.60×10^{-19} C
Electronvolts per atom	1 eV/atom = 23.05 kcal/mol
Erg	1 erg = 2.389×10^{-8} cal =1×10^{-7} J
Faraday	1 Faraday = 96,500 C = 23,070 call/V =
	$6.023 \times 10^{23} e^-$
Gas constant	R = 0.08205 liter atm/mol °K = 1.987
	cal/(mol °K) = 8.314×10^7 ergs/(mol°K)
Gram	1 g mass = 2.15×10^{10} kcal
Gram molecular volume	V_0= 22.413 liter/mol at 0°C, 1 atm
Grams per cubic centimeter	1 g/cm^3 = 62.43lb/ft^3
Ice point	T_0 = 273.15 K
Inch	1 in. =2.540 cm
Joule	1 J= 0.2390 cal
Kilogram	1 kg = 2.205 lb

Liter	1 liter = 1000 cm^3 = 1.0567 quarts (qt)
Liter-atmosphere	1 liter atm = 24.22 cal
Natural logarithms	ln x = 2.3026 log$_{10}x$
Mass of earth	M = 5.983 × 10^{24} kg
Pi	π = 3.1416
Planck's constant	h = 6.626 × 10^{-27} erg s
Pound	1 lb = 453.6 g
Quart	1 qt =0.9463 liter
Speed of light	c = 2.998 × 10^{10} cm/s
Speed of sound	v = 331.7 m/s (in air at STP) = 1470 m/sec (in water at 20°C)
Standard temperature and pressure	STP = 0°C/760 mm Hg
Stefan-Boltzmann constant	σ = 5.6705 × 10^{-5} erg/cm^2· s (K)^4

Appendix E
Measures and Equivalents

Length	Equivalent
1 millimeter	0.001 meter
	0.03937 inch
1 nanometer	1×10^{-9} meter
1 centimeter	10 millimeters
	0.3937 inch
1 inch	25.4 millimeters
1 decimeter	10 centimeters
	3.937 inches
1 foot	30.48 centimeters
	12 inches
1 meter	100 centimeters
	39.37 inches
1 yard	0.9144 meter
	3 feet
1 fathom	6 feet
1 dekameter	10 meters
	1.9884 rods
1 rod	0.5029 dekameter
	5.5 yards
1 furlong	40 rods
	0.125 miles
1 hectometer	100 meters
1 kilometer	1000 meters
	0.6214 miles
1 statute mile	1609.3 meters
	5280 feet
	8 furlongs
1 nautical mile	6076 feet
	1.15 statute miles
1 league	3 nautical miles
1 light year	9.4637×10^{9} meters
	5.8804×10^{12} miles

Area	Equivalent
1 square centimeter	100 square millimeters
	0.1550 square inch
1 square inch	6.4516 square centimeters

1 square decimeter	100 square centimeters
	0.1076 square foot
1 square foot	9.2903 square decimeters
	144 square inches
1 square meter	100 square decimeters
	1.196 square yards
1 square yard	0.8361 square meter
	9 square feet
1 square dekameter	100 square meters
1 square hectometer	100 square dekameters
1 square kilometer	100 square hectometers
	0.3861 square mile
1 square mile	640 acres
	2.59×10^6 square meters
1 acre	160 square rods
	43,560 square feet
	4046.9 square meters

Volume and Capacity	**Equivalent**
1 cubic centimeter	0.06102 cubic inch
1 cubic inch	0.0164 liter
1 cubic decimeter	0.0353 cubic foot
1 cubic foot	28.32 liters
1 cubic meter	1.308 cubic yards
1 liter	1,000 cubic centimeters
	61.024 cubic inches
	0.0353 cubic foot
	0.9081 quart dry
	0.0284 bushel
	1.0567 quarts liquid
	2.2046 pounds (lb) of water (4°C)
1 dekaliter	2.6417 gallons

Time	**Equivalent**
1 day	1440 minutes
	86,400 seconds
1 week	7 days
1 sidereal year	365.256 days
	8766.144 hours

Velocity	**Equivalent**
1 meter per second (m/s)	2.2369 miles per hour
1 mile per hour (mph)	1.4667 feet per second (ft/s)
	1.6093 kilometers per hour (km/h)
	0.447 meter per second (m/s)
1 knot	1 nautical mile per hour
	1.1508 statute miles per hour
1 radian per second	9.549 revolutions per minute
1 revolution per minute (rpm)	0.10472 radian per second

Angle	**Equivalent**
1 radian	$360/2\pi$ degrees
	57.296 degrees
1 degree	0.01745 radian
	60 minutes

1 minute	2.909×10^{-4} radian
	60 seconds
1 second	4.85×10^{-6} radian
1 mil (military)	360/6400 degree (°)
	0.05625 degree
	0.00098 radian
1 solid angle	4π steradians
	1 sphere
1 steradian	0.07958 solid angle

Weight	Equivalent
1 gram	0.03527 ounce
	15.4354 grains
1 kilogram	2.2046 pounds
1 quintal	100 kilograms
1 ton (English)	10 quintals
1 ton (metric)	1.1023 tons (English)
1 short ton (Avoirdupois)	2000 pounds
1 long ton (Avoirdupois)	2240 pounds

Energy and Power	Equivalent
1 horsepower	746 watts (W)
	33,000 ft lb/min
	2,544 Btu/h
1 horsepower-hour	273,740 kg meters
1 foot-pound	1356 J
	0.13826 kg meters
1 watt	1 J/s
	3.413 Btu/h
	44.22 ft lb/min
1 kilowatt	1.34 horsepower
	56.9 Btu/min
1 British thermal unit (Btu)	1055 Ws
	778 ft lb
1 joule	0.73756 ft lb

Acceleration	Equivalent
1 ft/s^2	30.480 cm/S^2
	0.6818 mi/h · s

Pressure	Equivalent
1 kg/cm^2	14.223 lb/in^2
	0.9678 normal atmosphere
1 kg/m^2	0.2048 lb/ft^2
1 lb/in.^{2}	6894.7 pascals
	27.71 in. of water
	2.04 in. of mercury
	0.06805 normal atmosphere
1 lb/ft^2	4.882 kg/m^2
1 in. of water	0.0361 lb/in.^{2}
	0.0735 in. of mercury
1 in. of mercury	0.4912 lb/in.^{2}
	13.58 in. of water

1 normal atmosphere	1.0133 bars
	1.0332 kg/cm^2
	14.696 lb/in.^{2}
	33.95 ft of water
	760 mm of mercury
1 pascal (Pa)	0.000145 lb/in.^{2}

Force	**Equivalent**
1 newton	1×10^5 dyn
	0.22481 lb
1 dyne (dyn)	2.2481×10^{-6} lb
	7.233×10^{-5} poundal
1 poundal	0.03108 lb
	1.3825×10^4 dyn

Torque	**Equivalent**
1 dyn · cm	1.0197×10^{-8} kg · m
	7.3757×10^{-8} ft lb
	2.3731×10^{-6} ft poundal

Temperature	**Equivalent**
Δ1 degree Centigrade (°C)	Δ1.8 degrees Fahrenheit (°F)
0°C	32°F
100°C	212°F

Appendix F
Basic Photometric Considerations

The subject of photometry deals with the quantities of visible light that occur at the scene being viewed or recorded and the corresponding light level at the system detector. This appendix contains information that will be helpful in understanding a few of the fundamental points involved. First is a brief review of the exposure value (E_v) system.

Basic Factors

Correct exposure results when the amount of light delivered to the image plane is in agreement with the light level required by the sensor to produce an optimum final image. There are just four contributing factors that must be considered in order to ensure that this balance has been achieved:

- the brightness of the scene
- the sensitivity of the sensor
- the $f/\#$ of the optics
- the time (duration) of the exposure.

The relationship between factors is demonstrated by the fact that the effect of either a brighter scene or a more sensitive film can be balanced by a shorter exposure time or a smaller aperture setting (larger $f/\#$). A more useful and detailed understanding of these relationships will result from a review of the E_v system.

E_v System

The E_v system was developed in the 1950s by German and American groups involved in scientific standardization. The basic information generated at that time has been updated, expanded, and is presented in tabular format in Fig. F.1.

A_v	f/#	T_v	Shutter Speed (Seconds)	S_v	ISO Rating	B_v	Scene Brightness (Ft L)	Scene Illuminance (Lux)	Scene Illuminance (Ft C)	Typical Exterior	Typical Interior
11	44	11	1/2000	11	6400	11	2000	120 K	11 K	Sunny Day	
10	32	10	1/1000	10	3200	10	1000	60 K	5600		
9	22	9	1/500	9	1600	9	512	30 K	2800		
8	16	8	1/250	8	800	8	256	15 K	1400	Open Shade	
7	11	7	1/125	7	400	7	128	8 K	740		
6	8	6	1/60	6	200	6	64	4 K	360		
5	5.6	5	1/30	5	100	5	32	2 K	180	Overcast Day	Well Lit Arena
4	4	4	1/16	4	50	4	16	1 K	90		
3	2.8	3	1/8	3	25	3	8	480	45		
2	2	2	1/4	2	12	2	4	240	22		
1	1.4	1	1/2	1	6	1	2	120	11		Daytime
0	1	0	1	0	3	0	1	60	5.6	Sunrise Sunset	
-1	.7	-1	2	-1	--	-1	.5	30	2.8		
-2	.5	-2	4	-2	--	-2	.25	15	1.4	Twilight	Nighttime
-3	-	-3	8	-3	--	-3	.12	7	0.6		

$$\text{Exposure Value} = E_v = A_v + T_v = S_v + B_v$$

Figure F.1 The exposure value (E_v) system.

Figure F.1 shows how each of the exposure factors, system $f/\#$ (A_v), shutter speed (T_v), sensor ISO rating (S_v), and scene brightness (B_v), can be assigned a value ranging from –3 to +11. Each incremental increase in that value represents a doubling of the factor's absolute value. The key to applying the information in Fig. F.1 is contained in the basic exposure value formula:

$$E_v = A_v + T_v = S_v + B_v.$$

From this formula we see that, for correct exposure, the sum of the system $f/\#$ and shutter speed values must be equal to the sum of the sensor ISO rating and scene brightness values.

Using the Table

The use of the E_v system and the tabulated information presented here is best demonstrated by applying it to two typical examples. For the first, assume that we are about to embark on a photographic excursion on a typical fall day. Since the camera will be handheld, a shutter speed of at least 1/60 of a second ($T_v = 6$) will be assumed. For best color rendition we select a fine-grain color slide film with an ASA rating of 25 ($S_v = 3$). As the sun passes in and out of clouds, the scene brightness will vary from overcast to bright sunny ($B_v = 8 \pm 2$). Plugging this data into the E_v equation, we have

$$E_v = A_v + T_v = S_v + B_v$$
$$A_v + 6 = 3 + 8(\pm 2)$$
$$A_v = 5(\pm 2).$$

From Fig. F.1 we can see that this value for A_v corresponds to lens aperture setting of $f/5.6$ on average, with a maximum of $f/11$ and a minimum of $f/2.8$, depending on the scene brightness. If it is felt that at $f/2.8$ the image quality, or depth of field, might not be ideal, we can improve on that situation by choosing a more sensitive film. Selection of a film with an ASA rating of 100 ($S_v = 5$), for example, will increase the E_v value by 2, resulting in a more desirable lens aperture range of from $f/5.6$ to 22.

For a second example, assume we wish to take photos at a sporting event in a well-lit indoor arena where the scene brightness will average about 16 foot-lamberts (fL) ($B_v = 4$). Selection of a zoom lens limits us to a maximum lens aperture of $f/4$ ($A_v = 4$). The fast-paced action would dictate a shutter speed of about 1/250 of a second ($T_v = 8$). This allows us to plug three of the four factors in the E_v equation and solve for the fourth as follows:

$$E_v = A_v + T_v = S_v + B_v$$
$$4 + 8 = S_v + 4$$
$$S_v = 8.$$

From the table, we find that an S_v value of 8 corresponds to a rather fast film with an ISO rating of 800.

Alternate Application

A basic understanding of the E_v system, in conjunction with the information given in Fig. F.1, makes it possible to utilize a handheld light meter, or a camera with built in light meter, to determine scene brightness or illuminance in absolute terms.

For example, suppose we wish to determine the light level that exists at our desktop. With a handheld light meter, set to an ASA setting of 100 ($S_v = 5$), a meter reading of reflected light can be taken by pointing the meter at the desktop. Assume that the meter indicates the scene should be photographed using a lens aperture of $f/4$ ($A_v = 4$), and a shutter speed of 1/30 of a second ($T_v = 5$).

Now, the scene (desktop) brightness (B_v) can be found

$$B_v = A_v + T_v - S_v$$
$$B_v = 4 + 5 - 5$$
$$B_v = 4.$$

From Fig. F.1 we find that a B_v of 4 corresponds to a brightness of 16 fL and (assuming a standard scene reflectance of 18%) a desktop illuminance of 960 lux, or 90 footcandles (fc).

Summary of the E_v System

Use of the standard exposure value formula and the data in Fig. F.1 leads to a better understanding of those factors involved in determining correct exposure. With this as a guide, it is possible to evaluate most imaging situations in advance and determine the best approach to achieve properly exposed images. While this method will yield reasonable estimates, best results will be achieved through the use of a modern light metering system to confirm these estimates and to make the ultimate determination of the correct exposure setting.

In an alternate application, we have seen that it is possible to dust off our rarely used light meter, or use our camera's built-in meter, in conjunction with the data presented here to determine absolute light conditions at a nearby or remote location.

Image Illuminance

Modern systems dealing with electronic imaging will frequently specify the performance of the system detector as a function of the illumination level that is present at that detector. The formula given in Figure F.2 shows the relationship that exists between illuminance at the scene and illuminance at the image (detector) plane. By applying this formula, it is possible to determine the range of lens speeds that must be provided by the system in order to image a designated range of scene brightness with a given detector. When it is found that the range of scene brightness is greater than can be accommodated by the system *f/#*, it is possible to modify the system transmission by the incorporation of switching, or variable, neutral-density filters.

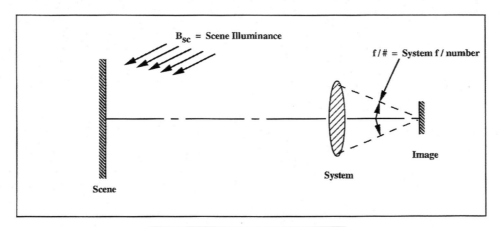

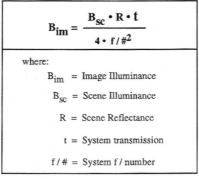

$$B_{im} = \frac{B_{sc} \cdot R \cdot t}{4 \cdot f/\#^2}$$

where:

B_{im} = Image Illuminance

B_{sc} = Scene Illuminance

R = Scene Reflectance

t = System transmission

$f/\#$ = System f/number

Figure F.2 Image illuminance.

Appendix G
Surface Sag and Conic Sections

Surface Sag (Sphere)

During the process of optical system design and layout, it is often necessary to determine the depth, or sag, of an optical surface at some specific height (aperture radius). For example, consider the configuration of a simple lens cell, as shown in Figure G.1(a). In order to protect the glass from physical damage by assuring that the lens vertex does not extend outside the lens cell, it is necessary to determine the sag of the convex lens surface at the height designated Y. The exact formula for the surface sag S is

$$S = R - (R^2 - Y^2)^{1/2}.$$

When the ratio of R to Y is greater than 5:1, the following formula can be used to calculate the approximate sag, with an error of less than 1%:

$$S \approx \frac{Y^2}{2R}.$$

Parabola

By far the most common shape for an optical surface is spherical. The spherical surface is created by rotating a circular section about the optical axis. Another surface shape that is frequently encountered in optical system design is the paraboloid. In this case a parabolic cross section is rotated about the optical axis to produce the solid paraboloidal surface. The paraboloid has the unique—and frequently very useful—characteristic of focusing all light rays in a collimated on-axis light bundle to a common point when the paraboloid is in the form of a concave reflector. In other words, the parabola is completely free of on-axis aberrations (spherical aberration and color).

In the case of the paraboloid, the formula given above for the approximate sag of a sphere becomes exact. For a parabolic surface,

$$S = \frac{Y^2}{2R}.$$

Figure G.1(b) shows the relationship between the parabola and the circle, regarding their respective sag formulas and the departure of the parabola from its reference (vertex) circle.

Other Conic Sections and the Conic Constant

Two other surface shapes are derived from cross sections taken through a solid right-angle cone: the ellipse and the hyperbola. Any of these conic sections can be described for purposes of optical analysis by designating their vertex curvature and conic constant.

Figure G.1(c) shows a plane intersecting a cone in four different orientations. In the upper left, the plane is parallel to the base of the cone and the resulting cross section is a circle. In the upper right, the plane is tipped downward and the resulting cross section is an ellipse. When the plane is tipped to the point where it becomes parallel with the far side of the cone (lower left), the cross section that results is a parabola. Finally, when the plane is tipped such that it is perpendicular to the base of the cone (lower right), the cross section that results is a hyperbola.

In most optical design software, the conic surfaces are described by designating their vertex radius along with a conic constant (*cc* or *k*). Fig. G.1(c) indicates the corresponding conic constant for each of the four common conic sections.

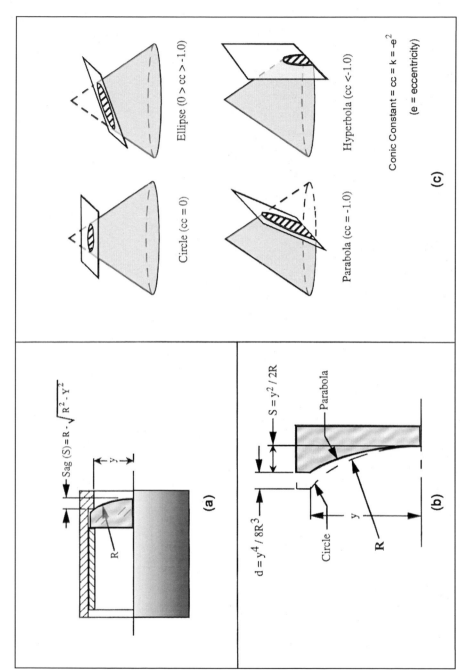

Figure G.1 Surface sag and conic sections: (a) basic sag *S* of a spherical surface, (b) departure *d* of a parabola from a circle, and (c) basic conic sections.

Index

distortion, 202, 223
  curve, 201
  negative, 112
  positive, 112
Dolland, John, 10
double-Gauss lens, 220
doublet, 123
  air-spaced, 124, 128, 129
  cemented, 124
dry nitrogen, 69
durability, 136
DVD (digital video disk), 237
  player, 241

Earth's atmosphere, 181
Eastern Optical Co., 16
Eastman Kodak, 14
Edmund
  Scientific, 91
  Optics Inc, 97
effective focal length (EFL), 47
electromagnetic spectrum, 21, 44
electronic imaging, 239
ellipse, 266
encircled fractional energy, 119
endoscope, 169, 236
ensquared energy diagram analysis, 223
ensquared fractional energy, 119
entrance pupil, 49, 202, 220, 221
equi-convex, 71
Eratosthenes, 6
erecting eyepiece, 166
Euclid, 5
exit pupil, 49, 153, 158, 162, 202
exposure factors, 261
exposure value (E_v) system, 259
eye
  convergence, 232
  relief, 162, 165, 202
eyelens, 189
eyepiece, 124, 152

$f/\#$, 56, 60, 156, 259, 261
farsightedness, 192
fiber optics, 170
  bundle, 172
field
  curvature, 108, 150
  curves, 200
  flattener, 112, 221
    lens, 223
  sag curves, 112
  stop, 50, 158, 165, 202
    size, 155
field of view, 47, 60, 153, 156, 158, 162, 163, 191
  observed, 150
  vertical, 165
final design optimization, 222
finest resolvable pattern, 147
Fizeau, Armand, 11
flint glass, 124, 175
focal system, 91
focus, 157, 235
  adjustment, 197
  range, 155
  shift, 139, 144
  travel, 198
Foucault, Jean Léon, 13
fovea, 191
fractional energy, 119
Fraunhofer, Joseph, 9
free aperture, 130
frequency, 24
front surface mirror, 136
fused
  quartz, 178
  silica, 173, 178

Galilei, Galileo, 8
Gaussian thin-lens formulas, 63
germanium, 182
Giambattista della Porto, 7
glare stops, 202